ACS SYMPOSIUM SERIES 297

Chromatography and Separation Chemistry

Advances and Developments

Satinder Ahuja, EDITOR
CIBA-GEIGY Corporation

Developed from a symposium sponsored by
the Division of Analytical Chemistry
at the 188th Meeting
of the American Chemical Society,
Philadelphia, Pennsylvania,
August 26–31, 1984

American Chemical Society, Washington, DC 1986

Library of Congress Cataloging-in-Publication Data

Chromatography and separation chemistry.
  ACS symposium series, ISSN 0097-6156; 297)

  "Developed from a symposium sponsored by the
Division of Analytical Chemistry at the 188th Meeting
of the American Chemical Society, Philadelphia,
Pennsylvania, August 26-31, 1984."

  Includes bibliographies and index.

  1. Chromatographic analysis—Congresses.
2. Separation (Technology)—Congresses.

  I. Ahuja, Satinder, 1933-   . II. American
Chemical Society. Division of Analytical Chemistry.
III. American Chemical Society. Meeting (188th: 1984:
Philadelphia, Pa.) IV. Title. V. Series.

QD79.C4C4838   1986   543′.089   85-28694
ISBN 0-8412-0953-7

ACS Symposium Series

M. Joan Comstock, *Series Editor*

Advisory Board

FOREWORD

The ACS SYMPOSIUM SERIES was founded in 1974 to provide a medium for publishing symposia quickly in book form. The format of the Series parallels that of the continuing ADVANCES IN CHEMISTRY SERIES except that, in order to save time, the papers are not typeset but are reproduced as they are submitted by the authors in camera-ready form. Papers are reviewed under the supervision of the Editors with the assistance of the Series Advisory Board and are selected to maintain the integrity of the symposia; however, verbatim reproductions of previously published papers are not accepted. Both reviews and reports of research are acceptable, because symposia may embrace both types of presentation.

CONTENTS

PREFACE

CHROMATOGRAPHY IS A POWERFUL SCIENCE concerned with separations of a large number of complex compounds. The contributions of chromatography to various scientific disciplines remain unmatched as is exemplified by the progress made in biological sciences such as biotechnology, clinical pharmacology/therapeutics, and toxicology. This book was planned to compile some of the exciting developments in the science of chromatography and separation chemistry.

A unified theory recently proposed to explain the manner of sorption and the form of sorption isotherm in gas, liquid, and ion-exchange chromatography is presented in some detail. Selectivity in reversed-phase high-pressure liquid chromatography is explored at length. Several chapters deal with characterization of bonded phases, relationship of column-packing structure and performance, variability of reversed-phase packing materials, and the differences between silica-based reversed-phase and poly(styrene–divinylbenzene) columns. A short review is included to cover various approaches used in HPLC to achieve the desired selectivity for resolution of enantiomeric compounds.

Detectors provide another route to optimizing selectivity by enhancing detectability of various classes of compounds. Discussion is included that describes the noise sources in optical detectors that allow optimization of such detectors. Two additional chapters deal with new and potential detectors including laser fluorimetry for HPLC, microcolumn HPLC, and FIA.

Effective use of computerization in chromatography is further advancing this science. Detailed information is provided regarding calculation and prediction of retention data for isocratic separations that allow optimization of methodologies. Calculation of retention data for complex gradient elution is also shown.

Recent developments in separation science are covered from the standpoint of their impact on various modes of chromatography. Chromatographic applications of cyclodextrin and phenylboronic acid stationary phases are discussed. The renaissance in electrophoresis has given rise to high-resolution, two-dimensional gel electrophoresis. The basic methodology and recent advances are provided for studies on protein separations. Finally, the combination of capillary supercritical fluid chromatography and mass spectrometry is discussed as another powerful tool for separation and detection of a variety of compounds for which GLC and HPLC cannot be used because of their apparent shortcomings.

SATINDER AHUJA
CIBA–GEIGY Corporation
Suffern, NY 10901

May 20, 1985

Manner of Sorption and Form of the Sorption Isotherm in Gas, Liquid, and Ion-Exchange Chromatography

R. J. Laub

Department of Chemistry, San Diego State University, San Diego, CA 92182

The Langmuir model of sorption and isotherm relation are
discussed in terms of approximations made in their formulation. An elemental modification of the latter is shown to yield
an equation that, in principle, accounts for adsorbate
asymmetry, surface heterogeneity, multilayer formation, and
sorption–desorption hysteresis and, hence, that encompasses the
BET model and isotherm Types 0 through V. Summation extension yields a comprehensive equation with which stepped
isotherms can also be represented. The form of the latter is
used to describe retentions in gas, liquid, and ion exchange
chromatography; the powers and coefficients of the resultant
expressions are shown in the first and third of these cases to be
rigorously interpretable in terms of the solute activity and
partition coefficients.

The method employed for choosing appropriate systems and conditions for a
particular analytical–scale chromatographic separation has traditionally
amounted to maximizing the column efficiency and, at the same time,
choosing (often nearly at random) stationary and/or mobile phases that are
thought to be suitable for the sample at hand (optimizing the system
selectivity). Maximizing the efficiency of chromatographic systems is taken
to be well understood at the present time, at least in principle (e.g., 1); even
so, while this method of effecting separations is straightforward and in fact
frequently provides a satisfactory solution, very many instances are
encountered wherein, if for no other than practical reasons, the selectivity of
the system must be enhanced as well. However, given the paucity of
quantitative guidelines regarding the latter, decisions made pertaining to
appropriate experimental conditions and technique(s) with which to achieve
this end have in the past inevitably been based primarily upon the experience
of the analyst as guided by what literature reports there might be that
describe ancillary analyses (e.g., 2).

An alternative approach to optimizing the system selectivity, and one
more rational than that of trial-and-error, would appear to be understanding,
first, of the physicochemical basis of chromatographic separations (3–7);
followed by modeling of the system at hand; thence optimizing the system and
conditions within the boundaries of whatever level of resolution is sought (8;

0097–6156/86/0297–0001$09.25/0
© 1986 American Chemical Society

see the reviews 9,10). However, while the third of these areas is now adequately characterized, the first and second remain only poorly understood since, presupposed, is a knowledge of the physicochemical interactions that take place between the solute and the mobile and stationary phases, as well as those that may occur between the mobile and stationary phases themselves. These may, in addition, be interdependent. Any model of the selectivity of the chromatographic process must therefore take account of such interactions; one such theory, founded upon a modified form of the Langmuir model of sorption, is formulated in what follows.

GENERALIZED REPRESENTATION OF ISOTHERM TYPES

The Langmuir Model

The pioneering consideration of isotherms, including their representation and interpretation, was carried out by Langmuir at the turn of this Century (11). His model is that of a symmetric adsorbate (solute) that interacts with an adsorbent comprised of S uniform sites such that the solute becomes attached to (adsorbs on) the solid surface. The process is said to result in S_1 sites of the adsorbent becoming occupied, leaving S_o sites unoccupied, where

$$S = S_o + S_1 \tag{1}$$

is presumed to obtain.

The rate at which sites are vacated (i.e., the rate of adsorbate evaporation from the surface) is said to be given by $k_1 S_1$, whereas the rate at which sites become occupied in the first place (i.e., the rate of adsorbate condensation) is $k_2 \, p_A S_o$, where k_i are rate constants, and where p_A is the partial pressure of adsorbate above the adsorbent. Thus, at equilibrium,

$$k_1 S_1 = k_2 p_A S_o = k_2 p_A (S - S_1) \tag{2}$$

Now let Θ be the fraction of sites occupied at any time:

$$\Theta = S_1/S \tag{3}$$

Then, upon substituting equation 2 into 3, followed by rearrangements,

$$\Theta = \frac{k_2 \, p_A}{k_2 p_A + k_1} \tag{4}$$

Letting $b = k_2/k_1$, the ratio of the rate constants (i.e., an equilibrium constant) which is taken to be independent of p_A,

$$\Theta = \frac{b \, p_A}{1 + b \, p_A} \tag{5}$$

Alternatively, Θ can be replaced by the volume ratio v_A/V_A, where v_A is the volume of adsorbate actually on the surface, and where V_A is the volume of adsorbate corresponding to monolayer coverage: $\Theta = v_A/V_A$. Also, since the density of the adsorbate is assumed to be unaffected by the adsorption process, Θ could equally be represented by a mole or weight ratio.

The denominator of equation 5 tends to unity at very low pressure, whence $v_A = b V_A p_A$; plots of v_A against p_A are then predicted to be linear and of zero intercept. When in contrast $(b p_A)$ is not negligible in comparison with unity, the plots are predicted to be curved concave to the abscissa, i.e., the well-known Langmuir isotherm.

The BET Model

In practice, isotherm plots are of course found to describe a number of forms different from those mentioned above. Each has become known as a Type, where Type 0 is the ideal (i.e., linear) case, followed by Type I, the Langmuir isotherm. Examples of Types II–VI are also known, Brunauer, Emmett, and Teller (12) being the first to collate systems that exhibit such curve shapes.

These workers were also the first to model multilayer sorption: the Langmuir assumption of a uniform surface is retained; further, solutes in the first layer are said to be localized to (i.e., immobilized on) a given site. Additional adsorbate molecules are then permitted to stack (but not interact) in layers on top of one another, where molecules in the second and subsequent layers are taken to have properties approximating those of bulk condensate. The resultant formulation with which isotherms of at least through Type V can be reproduced is then given by:

$$\Theta = \frac{(c/p_A^o)\; p_A}{(1 - p_A/p_A^o)\left[1 + (c - 1)p_A/p_A^o\right]} \tag{6}$$

where c, a dimensionless constant greater than or equal to unity, is approximately $\exp(H_1 - H_L)/RT$, that is, reflects the (temperature-dependent) difference in the heats of sorption of the first and subsequent layers of adsorbate; and where p_A^o is the bulk-solute vapor pressure.

[The more familiar form of the BET equation can be obtained by substitution of v_A/V_A for Θ, followed by rearrangements:

$$\frac{p_A/p_A^o}{v_A(1 - p_A/p_A^o)} = \frac{(c - 1)p_A/p_A^o}{V_A c} + \frac{1}{V_A c}$$

where plots of the left-hand side of the relation against p_A/p_A^o are predicted to be linear and from which both V_A and c can be evaluated, the former from the inverse of the sum of the slope and intercept, and the latter from 1 + the slope/intercept quotient.]

Extension of the Langmuir Model to Include Equation 6

As a prelude to application of the Langmuir formalism to chromatography, we first enquire whether some or other simple modification of it can accomodate the BET relation, equation 6, since, if so, the resultant generalized expression might then be indicative of an appropriate model of the sorption process relevant to analytical separations. Accordingly, we rewrite the Langmuir equation with b replaced by a second constant b' in the denominator of equation 5:

$$\Theta = \frac{b \; p_A}{1 + b' \; p_A} \tag{7}$$

where $b = c/p_A^o$, and where setting $(1 + b' \; p_A)$ equal to the denominator of equation 6 provides the identity:

$$b' = \frac{c - 2}{p_A^o} + p_A \left[\frac{1 - c}{(p_A^o)^2} \right] \tag{8}$$

whence b' is identified as being pressure-dependent.

The BET model is well-known to hold only in the limit of low pressure, so that equation 8 reduces approximately to:

$$b' \approx \frac{c - 2}{p_A^o} \tag{9}$$

When $c \gg 2$, $b' \approx c/p_A^o = b$, whence equation 7 further reduces to equation 5, i.e., the Langmuir model, with which isotherm Type I is described. In the limit of sufficiently low pressure, the relation describes Type 0 as well, as discussed above. In those instances where $c \approx 2$, $b' \approx 0$ which results also in a Type 0 isotherm. When p_A cannot be neglected then, for $c \gg 2$, isotherm Type II results, with the eventual appearance of the Type IV extension. In contrast, for $c \approx 2$, $b' \approx - p_A/(p_A^o)^2$ which yields isotherm Type III and, as p_A is increased, Type V. Equation 7 thereby accounts for the seemingly-inexplicable fact that stepped isotherms such as Types IV and V (the former being particularly common) are frequently observed at pressures far below that corresponding to saturation. (It is of interest in this regard that isotherm Type II is a composite of Types I and III, whereas isotherm Type IV is a composite of Types I + I. Similarly, isotherm Type V is comprised of Type III with extension to Type I.)

Comprehensive Isotherm Relation

The above derivations have been brought out in full in order to reveal the approximations made in each and, moreover, to illustrate that neither the Langmuir nor BET models account for multi-stepped isotherms despite modification of the form of equation 5. First, of course, adsorbate species of interest generally, let alone those in chromatography, self-evidently will rarely be of spherical symmetry. Thus, adsorbates can adsorb onto an adsorbent in different ways, for example, the hypothetical edgewise vs. planar adsorption of benzene onto a presumed-flat surface. In addition, not only can there be different conformations of the first layer of adsorbate on an adsorbent, but this layer may also orient the second layer, which in turn can orient the third layer, and so on. The phenomenon is well-known in the sorption of liquid-crystalline materials onto solid surfaces, the so-called "boundary layer" (that is, that which is intermediate between the surface and bulk liquid) being known to extend in some cases over thousands of Angstroms (13). However, the ordering is not necessarily unidirectional (nematic-like) and can in fact extend in two dimensions, analogous to two-dimensional ordering of smectic liquid crystals. (With chiral adsorbates, there could evidently be superposed a twist as well.)

The second major objection to the Langmuir and BET formulations derives from consideration of the adsorbent surface. Both models assume a finite number of uniform sites available for adsorption, but even cursory microscopic evaluation of surfaces of interest in chromatography demonstrates that with the possible exception of smooth glass beads, this is rarely the case. Surfaces such as diatomaceous earth, silica, etc. are highly heterogeneous and, in addition, possess microporous structure, the adsorptive properties of which can be much different from those of the surface (14).

Also implicit in the models is that the surface remains unchanged when contacted with an adsorbate. This is now known to be incorrect in at least several instances, and adsorbates can in fact cause a reorganization of adsorbent surface atoms (15).

Finally, because adsorbate molecules are adsorbed in a specific orientation (or array of orientations), there will be hysteresis involved in the sorption/desorption process (16). Hysteresis is known to arise also as a result of adsorbent-site heterogeneities (17).

In order to take account of all of the above effects, we rewrite equation 7 at this point as a summation:

$$ \Theta = \sum_{i=1}^{n} \left[\frac{b_i \, p_A}{1 + b_i' \, p_A} \right] \tag{10} $$

where isotherm Types 0-V can be represented as described above with n = 1, while multi-stepped isotherm shapes require that $n > 2$ with $b_i \neq b_i'$. The alternative (but equivalent) view is a virial expansion of equation 8, the first three terms of which must evidently be:

$$ b' = -\frac{1}{p_A^o} + \frac{1}{p_A^o}(c - 1) - \frac{p_A}{(p_A^o)^2}(c - 1) + \dots \tag{11} $$

Multi-stepped isotherm shapes can therefore be described either by expansion, as required, of the summation of equation 10, or by virial extension of equation 11; while hysteresis is taken into account by replacing b_i in the denominator of equations 5 et seq. with b_i'. Moreover, in doing so, and in contrast to the BET model, each of the second and higher layers of adsorbate can be distinguished as is important, for example, in instances of liquid-crystalline solutes.

APPLICATION OF EQUATION 10 TO GAS CHROMATOGRAPHY

Generalized Representation of Retentions with Blended Sorbents

It has been shown in very many (but by no means all) instances of the use of binary stationary phases in gas chromatography, that solute partition coefficients at hypothetical zero column pressure drop K_R^o (i.e., corrected for virial effects; 18) are described by the relation (19-23):

$$K_{R(M)}^{o} = \phi_B \, K_{R(B)}^{o} + \phi_C \, K_{R(C)}^{o} \tag{12}$$

where the subscripts refer to pure stationary phases B or C or to a mixture of these (M = B + C), and where ϕ_i (i = B or C) is a volume fraction. In addition, Laub has described and discussed various modifications that can be incorporated into equation 12 to take account of solute-solvent stoichiometric or nonstoichiometric complexation along with concomitant solvent-solvent self-association and/or solvent-additive interactions (24; see also ref. 25).

In order to assess whether the isotherm formalism might be applied to these situations, we write equation 12, first, in the form:

$$K_{R(M)}^{o} = \phi_C \left\{ K_{R(C)}^{o} + \sum_{i=1}^{n} \left[\frac{b_i \, \phi_B}{1 + b_i' \, \phi_B} \right] \right\} + \phi_B \, K_{R(B)}^{o} \tag{13}$$

We then seek to interpret the constants b_i, b_i' in what follows.

Non-Interactive Solute Solution with Non-Interactive Solvents

In the trivial case of what might well be referred to as elemental solute solution in the stationary-phase blend M, there is presumed to be neither solute specific chemical interaction with B or C, nor stoichiometric or non-stoichiometric solvent-solvent chemical interaction. Thus, allowance is made for nonspecific solute interactions with M (such solutions being designated variously as ideal, athermal, "regular", "solvated", random contact-paired, and so forth, depending upon the enthalpy and entropy of mixing), while the solvent-solvent solution is taken to be ideal. The regression of solute partition coefficients against the solution composition in these instances will be linear, and will accordingly be described by equation 13 with b_i, b_i' set equal to zero, i.e., equation 12. Since the pure-phase partition coefficients will in all likelihood differ, the slopes of such plots may be positive, zero, or negative, examples of all of which have been documented and discussed (19-21).

We note also that solvent-solvent "ideality" is mimicked by systems that exhibit partial miscibility. Thus, equation 12 is expected to hold in these instances as well, since the compositions of each phase of such two-phase systems are invariant with the notional concentrations of B and C. In testing this, Laub, Purnell, and Summers (26) evaluated the partition coefficients of several solutes with the solvent systems tributyl phosphate/ethylene glycol and ethyl benzoate/propylene glycol (upper consolute temperatures of 53° and 47°C, respectively), and found that not only was equation 12 obeyed as expected within the two-phase regions of each but also, that the lines were contiguous with those corresponding to the single-phase portions of the phase diagrams.

[As one consequence of the "ideality" of partially-miscible solvent systems, the use of mechanical blends of pure-phase packings is favored in packed-column analytical separations since, in the absence of any substantial volatility of B and C, equation 12 must be obeyed exactly and, hence, retentions with blended phases can be forecast from those with the two pure solvents (8,27; see the reviews 9,10,22). The same holds also for the use of tandem-connected packed (28,29) as well as capillary columns (30,31), provided that the pressure drop across each is correctly taken account of (32,33).]

Interactive–Solute Solution with Non–Interactive Solvents

In the simplest of these situations, solute complexation with one of the stationary-phase components, say C, in admixture with presumed–inert (i.e., non–interactive) diluent phase B, the Gil-Av/Herling relation (34) is substituted into equation 12 to provide:

$$K^o_{R(M)} = K^o_{R(B)} + \frac{K^o_{R(B)} \; K_{AC} \; \phi_C}{\bar{V}_C} \tag{14a}$$

$$= \phi_C \left[K^{o,t}_{R(C)} + \frac{K^{o,t}_{R(C)} \; K_{AC}}{\bar{V}_C} \right] + \phi_B \; K^o_{R(B)} \tag{14b}$$

where K_{AC} is the notional solute (A) complexation constant with C,

$$K_{AC} = \frac{\bar{V}_C \left[K^o_{R(M)} - K^o_{R(B)} \right]}{\phi_C \; K^o_{R(B)}} \tag{15a}$$

which, for $K^o_{R(M)}$ linear in ϕ [as in the several systems considered for example by Martire and Riedl (35); see also the many systems cited in refs. 19–21], reduces to:

$$K_{AC} = \frac{\bar{V}_C \left[K^o_{R(C)} - K^o_{R(B)} \right]}{K^o_{R(B)}} \tag{15b}$$

where the true (i.e., uncomplexed) partition coefficient of A with C, $K^{o,t}_{R(C)}$, is defined by:

$$K^o_{R(C)} = K^{o,t}_{R(C)} + \frac{K^{o,t}_{R(C)} \; K_{AC}}{\bar{V}_C} \tag{15c}$$

and where $\bar{V}_C$ is the molar volume of C.

Setting equations 15a and 15b equal immediately yields equation 12 (the manner, in fact, in which it was first derived), while substitution of equation 15c into equation 12 then provides equations 14.

Solute complexation with one of the solvent components, here C, is also described by the isotherm relation, equation 13, with n = 1:

$$K^o_{R(M)} = \phi_C \left[K^{o,t}_{R(C)} + \frac{b_1 \phi_B}{1 + b'_1 \phi_B} \right] + \phi_B K^o_{R(B)} \tag{16}$$

where equation 16 reduces to equation 14b (hence 14a) upon specification of the following identities:

$$b_1 = \frac{K^{o,t}_{R(C)} K_{AC}}{\phi_B} \tag{17a}$$

$$b'_1 = \frac{\bar{V}_C - 1}{\phi_B} \tag{17b}$$

In instances in which the solute interacts stoichiometrically both with B and C, equation 14b is expanded to:

$$K^o_{R(M)} = \phi_C \left[K^{o,t}_{R(C)} + \frac{K^{o,t}_{R(C)} K_{AC}}{\bar{V}_C} \right] + \phi_B \left[K^{o,t}_{R(B)} + \frac{K^{o,t}_{R(B)} K_{AB}}{\bar{V}_B} \right] \tag{18}$$

In terms of the isotherm formalism this is:

$$K^o_{R(M)} = \phi_C \left[K^{o,t}_{R(C)} + \frac{b_1 \phi_B}{1 + b'_1 \phi_B} + \frac{b_2 \phi_B}{1 + b'_2 \phi_B} \right] + \phi_B K^{o,t}_{R(B)} \tag{19}$$

i.e., equation 13 with n = 2, where in addition to b_1 and b'_1 given as above, b_2 and b'_2 are identitified as:

$$b_2 = \frac{K^{o,t}_{R(B)} K_{AB}}{\phi_C} \tag{20a}$$

$$b'_2 = \frac{\bar{V}_B - 1}{\phi_B} \tag{20b}$$

Equations 16 and 19 are indicative of the versatility of the form of the isotherm relation, equation 13, insofar as the latter provides for description of solute complexation with either (or both) of the solvent components of binary stationary phases. In addition, each of what might otherwise appear simply to be empirical fitting constants can be precisely interpreted (equations 17 and

20). Moreover, the relations can obviously be expanded to take account of multicomponent phases to the same extent that the diachoric solutions relation, equation 12, can be extended (36):

$$K^o_{R(M)} = \sum_{i=1}^{n} \phi_i \, K^o_{R(i)}$$

Equation 12 takes account as well of solute/stationary-phase partitioning in the instance of gas-solid chromatography since, in the absence of adsorbent-adsorbent interactions, (adsorbate) solute specific retention volumes are described exactly by the weight-average of the retentions with the pure adsorbents (37). This holds also in the instance of adsorbate/adsorbent charge-transfer complexation, i.e., equations 14 and 15 (38).

However, for binary-component liquid (or solid or interfacial) sorbent phases, plots of $K^o_{R(M)}$ against ϕ are predicted to be linear irrespective of whether the solute complexes with B (K_{AB} finite; K_{AC} zero), with C (K_{AC} finite; K_{AB} zero), or with neither (K_{AB}, K_{AC} each zero) or both (K_{AB}, K_{AC} each finite). That is, even though these various situations can be interpreted precisely in terms of the equations presented thus far, the isotherm formalism appears at this point to provide little advantage over the diachoric solutions equation in the speciation of solute complexation with non-interactive solvents.

Non-Interactive Solute Solution with Interactive Solvents

The two solvent components have thus far been assumed to be non-interactive, that is, to form ideal solutions. Since these are equivalent to those comprised of immiscible liquids, Laub and Purnell christened such systems as "diachoric" (partitioned volume) (20). That is, diachoric solutions are defined as those for which the solvent-solvent activity coefficients γ are unity and, hence, for which plots of solute partition coefficients against solvent composition must regress linearly.

For example, Martire and Riedl proposed some years ago that K_{AC} is calculable solely from the end-points of what amount to (straight-line) plots of equation 14a (i.e., equation 15b), provided that the two solvents are identical in all respects (yield ideal solutions) save that one or the other complexes with the solute (35). Such systems include chloroform with the stationary phases n-heptadecane + di-n-octyl ether, for which the activity coefficient $\gamma_{i(M)}$ of either solvent component (i = B or C) in the blend M (= B + C) was said to deviate from unity by at worst $\pm 0.00_7$ at $\phi = 0.5$ (where the observed K_{AC} was taken to reflect hydrogen bonding of the solute with the latter solvent).

When in contrast the solvent components somehow interact attractively or repulsively (i.e., exhibit negative or positive deviations from Raoult's law), the regression of $K^o_{R(M)}$ must be presumed to be affected also. However, the form that this might take has yet to be specified in any detail. Nevertheless, and in order that equations 12 et seq. might be applied to combinations of stationary phases that are of interest in solutions theories in general, and in analytical gas-liquid chromatography in particular, the effect of solvent-solvent interaction on solute/stationary-phase partitioning must be taken into account.

Modification of Equation 12 to Account for Solvent-Solvent Interaction. In every situation in which deviations from equation 12 have been determined accurately, the signs and magnitudes of the deviations have been found to be virtually identical for non-interactive solutes (e.g., alkanes) with a particular pair of solvents. For example, n-pentane through n-octane, cyclohexane, and methylcyclohexane yield partition coefficients with equimolar mixtures of squalane + dinonyl phthalate phases at 30°C that, uniformly, are ca. 7% greater than those predicted via equation 12 (39). Moreover, all exhibit (curvilinear) plots of $K_{R(M)}^{o}$ against ϕ_{DNP} that are slightly asymmetric toward the squalane ordinate. In contrast, the same n-alkane solutes deviate negatively and symmetrically from linearity by ca. 1% with n-octadecane + n-hexatriacontane at 80°C (40). Chien and Laub (41) have also shown that the deviations from equation 12 exhibited by n-pentane, n-hexane, and cyclohexane with the three combinations of n-hexadecane, n-hexadecyl iodide, and tetra-n-butyltin fall between these extremes, amounting to ca. 1-2% symmetric positive deviation at $\phi = 0.5$.

We are accordingly led to postulate that deviations exhibited by such non-interactive solutes from one or another forms of the diachoric solutions relation are in fact a reflection of deviations of the solvent components themselves from Raoult's law. We therefore define an arithmetic-mean activity coefficient $\bar{\gamma}_{M}$ for blends of B + C, to which we assign the form:

$$\ln \bar{\gamma}_{M} = \frac{g_{M}^{E}}{2RT} = \frac{1}{2}\left[x_{B} \ln \gamma_{B(M)} + x_{C} \ln \gamma_{C(M)}\right] \qquad (21)$$

where g_{M}^{E} is the excess free energy of mixing of B and C, and where x_{i} is a mole fraction. Referring $\bar{\gamma}_{M}$ to the behavior of non-interactive ($K_{AB} = K_{AC} = 0$) solutes with interactive solvents then leads to the expression:

$$K_{R(M)}^{o} = \left[\phi_{C} \Delta K_{R}^{o} + K_{R(B)}^{o}\right]\bar{\gamma}_{M} \qquad (22)$$

where $\Delta K_{R}^{o} = K_{R(C)}^{o} - K_{R(B)}^{o}$. Alternatively, the situation may be viewed from the standpoint that deviations from equation 12 yield the mean deviation of the solvent components from ideality. In either event, equation 22 can be tested immediately, provided that appropriate solvent/solvent activity-coefficient or free-energy data are available or can be calculated.

Solvent-Component Negative Deviation from Raoult's Law. A blended-solvent system that exhibits large negative deviations from Raoult's law is aniline (AN) + phenol (PH): the phase diagram (42) indicates that 1:1 AN:PH compound formation is extant up to the congruent melting point of 31°C (x = 0.5), with eutectic points occurring at 15° ($x_{AN} = 0.2$) and at -12°C ($x_{AN} = 0.9$). Further, Vernier, Raimbault, and Renon (43) have reported (glc-based) infinite-dilution liquid-gas activity coefficients $\gamma_{A(M)}^{\infty}$ for several solutes with this solvent system, from which $K_{R(M)}^{o}$ can easily be calculated. Accordingly, and following assessment of the excess free energy of mixing of the two solvents as a function of solution composition, a mean activity coefficient can be calculated utilizing equation 21. The solute partition coefficients can then be predicted via equation 22, comparison of which with those determined experimentally will provide a test of the consistency of the relations, as well as a measure of the extent to which deviations from the diachoric solutions expression can be attributed to solvent-solvent interaction.

To begin, Vernier and coworkers reported that the excess free energy of mixing of AN + PH at 34°C could be represented across the entire composition range to a fair degree of approximation by the Redlich-Kister equation:

$$g_M^E = RTx_{PH}x_{AN}\left[B' + C(x_{PH} - x_{AN}) + D(x_{PH} - x_{AN})^2\right] \quad (23)$$

where B', C, and D are empirical fitting constants, which were found to be −1.7112, −0.2650, and 0.7156, respectively. (For example, the excess free energy of mixing for the system at x = 0.5 was said to be −261 cal mol^{-1}.) We show the g_M^E calculated with equation 23 plotted in Figure 1 as open triangles. The data exhibit a point of inflection at x_{AN} = 0.85 and so, might be regarded as circumspect. Alternatively, a smooth curve can be generated with the (symmetric) one-term Margules expression:

$$g_M^E = A' R T \, x_{PH} \, x_{AN} \quad (24a)$$

That calculated with (A' R T) set equal to −950 cal mol^{-1} is shown plotted with open circles in Figure 1. However, the curves generated with equation 23 and equation 24a are both symmetric with respect to x_{AN}, which is difficult to reconcile with the diverse chemical nature of aniline and phenol (the values are being reassessed at the present time by differential scanning calorimetry, the results of which we intend reporting elsewhere). Thus, the excess free energy would seem to be better represented by the two-term (asymmetric) Margules expression:

$$g_M^E = R T (A \, x_{PH} + B \, x_{AN}) \, x_{PH} \, x_{AN} \quad (24b)$$

where the values: A = −0.8713, and B = −2.0879 (closed circles in Figure 1) were found to yield the best fit to the averages of the deviations from equation 12 of the partition-coefficient data for the alkane solutes (n-pentane, n-hexane, and n-heptane; crosses in Figure 1).

The latter result is clearly quite acceptable, the sole exception being the point at x = 0.5 (cross in brackets). This datum is consistent with 1:1 compound formation of the solvent components, which is reflected also by the anomalous solution expansion at 30°C reported for this composition (44). Thus, since the partition-coefficient data were taken at 34°C, i.e., at only 3° above the congruent melting point of the aniline:phenol complex, the deviation might well be due to the onset of stoichiometric solvent-solvent sociation. Moreover, we have observed similar anomalies in plots of partition coefficients against volume fraction at as much as 20° above eutectic points of mixed liquid-crystalline phases (45).

The self-consistency of equations 21, 22, and 24b is shown tested further in Figure 2, where the calculated (open circles) and experimental (filled circles) solute partition coefficients are plotted against ϕ_{AN}. The latter were calculated from the solute activity coefficients reported by Vernier and coworkers as follows. First, stationary-phase mole fractions were converted to volume fractions via the relation:

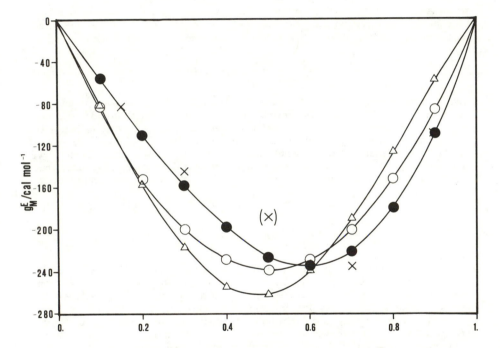

Figure 1. Plots of the excess free energy of mixing g_M^E of aniline + phenol at 34°C, calculated with equations 23 (triangles), 24a (open circles), and 24b (closed circles). Crosses: g_M^E calculated with equations 21 and 22 for n-alkane solutes.

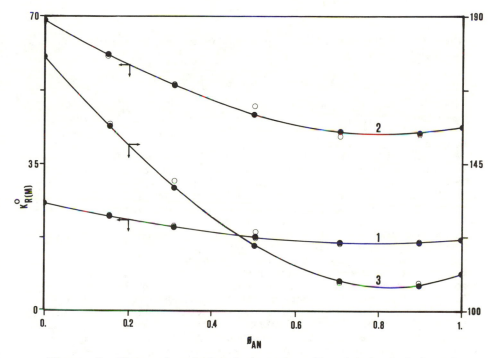

Figure 2. Plots of calculated (equation 22; closed circles) and experimental (open circles) partition coefficients of n-alkane solutes (1: n-pentane; 2: n-hexane; 3: n-heptane, right-hand ordinate) with aniline + phenol solvents at 34°C.

$$\phi_{AN} = \frac{x_{AN}}{x_{AN}(1-r)+r}$$

where r is the ratio of molar volumes of phenol (88.476 cm^3 mol^{-1}) and aniline (92.204 cm^3 mol^{-1}) (46), and where the excess volume of mixing is less than 0.7% at x = 0.5 (44). The experimental $K^o_{R(M)}$ were then deduced from the reported solute activity coefficients and the relation (cf. ref. 4):

$$K^o_{R(M)} = \frac{RT}{\gamma^\infty_{A(M)} \; P^o_A \; \bar{V}_M}$$

where the vapor-pressure data were obtained from ref. 46, and where $\bar{V}_M$ was calculated as usual from:

$$\bar{V}_M = x_{PH} \bar{V}_{PH} + x_{AN} \bar{V}_{AN}$$

The predicted $K^o_{R(M)}$, through which were constructed the solid curves, were calculated from the end-point partition coefficients, the $\bar{\gamma}_M$ derived from equations 21 and 24b, and equation 22.

The fits illustrated in Figure 2 are encouraging: the experimental partition coefficients conform everywhere to the calculated values to within error limits of at worst ± 3% except at ϕ_{AN} = 0.5, which we take as indicative of the onset of compound formation as noted above. Moreover, random scatter in the reported solute activity-coefficient data indicates that what deviations remain may well be due simply to experimental difficulties in using these volatile solvents as glc stationary phases (each exhibits a bulk vapor pressure of in excess of 1 torr at 34°C: 46).

In any event, and to the extent that the postulate can be tested with the data at hand, superposition upon the solute partition coefficients of the solvent-component deviations from Raoult's law, equation 22, appears to go far in accounting for deviations from the diachoric solutions relation. That is, the non-interactive probe-solute partition coefficients reflect interactions that are extant between the solvents, negative deviations from equation 12 being indicative (as here) of solvent-solvent activity coefficients of less than unity. Moreover, even if this were in fact to prove to be only approximately so, the way would nonetheless be open to a very powerful method of assessment (to the same degree of approximation) of (finite-concentration) solvent/solvent activity coefficients: one need only measure (infinite-dilution) probe-solute partition coefficients; then deduce $\bar{\gamma}_M$ via equation 22, thence g^E_M as well as $\gamma_{i(M)}$. An example of this is presented in what follows.

Solvent–Component Positive Deviation from Raoult's Law. The majority of blended-solvent systems studied by glc appear to yield solute partition coefficients that correspond to positive deviations from equation 12. Accordingly, we assert that such solvent pairs also exhibit positive deviations from Raoult's law. Unfortunately, however, there are no data to assess this at the present time. (The system di-n-octylmethanamine/n-octadecane is currently under study in this Laboratory; preliminary results suggest that $\bar{\gamma}_M$ = 1.7 at x = 0.5.) Nevertheless, we can at least examine the self-consistency of equation 22 by calculation of $\bar{\gamma}_M$ as outlined above for several non-interactive solutes with a given pair of stationary phases. Precise (47) and accurate (48) partition coefficients are available for a number of such species with squalane (SQ) + dinonyl phthalate (3,5,5-trimethylhexyl phthalate; DNP) (39) and so, this system was chosen for study in this portion of the work.

The experimental partition coefficients for n-pentane through n-octane with mixtures of SQ + DNP are presented below in Table I. The $\bar{\gamma}_M$ derived for the n-alkanes from the end-point partition coefficients, the data of Table I, and equation 22 are then presented in Table II.

The goodness of fit of the average $\bar{\gamma}_M$ for the n-alkanes is illustrated graphically in Figure 3, where the deviations $\Delta/\%$ of the calculated $K^o_{R(M)}$ relative to the experimental values are shown plotted against ϕ_{DNP} (closed circles: n-pentane; open circles: n-hexane; open triangles: n-heptane; open squares: n-octane). Also shown as horizontal dashed lines are the limits of overall experimental error for the particular apparatus and techniques employed in the work, which were estimated elsewhere (47) to be at worst ± 1.4% (the major contribution to which being determination of the column volume of liquid phase). Of the 40 data points, 15 lie above the abscissa, while 25 fall below. Further, only 2 points fall outside the error limits, these being for the most volatile solute, n-pentane, at high DNP content of the stationary phase. Nevertheless, reference to Table II indicates that there is what appears to be a systematic variation of $\bar{\gamma}_M$ as a function of carbon number. At ϕ_{DNP} = 0.5575, for example, the derived mean solvent activity coefficients are 1.045 (n-pentane), 1.056 (n-hexane), 1.063 (n-heptane)and 1.071 (n-octane). The standard deviation of the average is ± 0.011, however, and so, what systematic deviations there might be appear to correspond only to ca. ± 1%.

As a further test of the veracity of the derived $\bar{\gamma}_M$, we compare in Table III the experimental partition coefficients of cyclohexane and methylcyclohexane with the $K^o_{R(M)}$ calculated from the mean solvent activity

Table I. Experimental Infinite–Dilution Liquid–Gas Partition Coefficients of Indicated Solutes with Listed Volume Fractions ϕ_{DNP} of Squalane (SQ) + Dinonyl Phthalate (DNP) at 30°C

ϕ_{DNP}	$K^o_{R(M)}$			
	n-Pentane	n-Hexane	n-Heptane	n-Octane
0.	98.54	304.4	926.9	2791.
0.0751	96.30	299.1	909.1	2743.
0.1483	95.09	295.3	898.9	2695.
0.2153	93.74	291.4	886.3	2668.
0.2716	93.56	290.6	883.0	2633.
0.3252	92.52	285.3	864.6	2600.
0.4516	88.76	274.1	826.9	2493.
0.5575	84.01	258.2	779.3	2329.
0.7113	77.39	237.6	711.0	2105.
0.8200	73.71	224.5	664.5	1957.
0.9072	70.48	214.8	636.1	1864.
1.	66.05	197.1	578.9	1685.

Table II. Mean Solvent Activity Coefficients $\bar{\gamma}_M$ at Indicated Volume Fractions ϕ_{DNP} of Squalane (SQ) + Dinonyl Phthalate (DNP) at 30°C, Calculated from End-Point n-Alkane Solute Partition Coefficients of Table I and Equation 22

| ϕ_{DNP} | $\bar{\gamma}_M$ | | | | |
	n-Pentane	n-Hexane	n-Heptane	n-Octane	Ave.
0.0751	1.002_1	1.009_3	1.009_3	1.013_0	$1.008_4 \pm 0.004_6$
0.1483	1.014_6	1.023_6	1.027_0	1.025_9	$1.022_8 \pm 0.005_6$
0.2153	1.024_0	1.035_9	1.040_3	1.045_1	$1.036_3 \pm 0.009_0$
0.2716	1.042_9	1.055_7	1.060_8	1.057_2	$1.054_2 \pm 0.007_8$
0.3252	1.051_7	1.058_6	1.062_5	1.069_4	$1.060_6 \pm 0.007_4$
0.4516	1.058_3	1.070_9	1.074_3	1.087_9	$1.072_9 \pm 0.012_2$
0.5575	1.044_6	1.055_7	1.063_3	1.071_1	$1.058_7 \pm 0.011_3$
0.7113	1.026_0	1.041_8	1.046_6	1.056_2	$1.042_7 \pm 0.012_6$
0.8200	1.025_2	1.037_4	1.035_8	1.038_7	$1.034_3 \pm 0.006_2$
0.9072	1.020_5	1.037_4	1.040_8	1.042_7	$1.035_4 \pm 0.010_1$

Table III. Comparison of Experimental Liquid–Gas Partition Coefficients of Cyclohexane and Methylcyclohexane Solutes with $K_{R\,(M)}^{o}$ Calculated from Mean Solvent Activity Coefficient Data of Table II and Equation 22

| ϕ_{DNP} | $K_{R\,(M)}^{o}$ | | | |
| | Cyclohexane | | Methylcyclohexane | |
	Calc.	Exptl.	Calc.	Exptl.
0.	579.2	579.2	1166.	1166.
0.0751	569.6	568.5	1144.	1137.
0.1483	563.4	560.7	1129.	1124.
0.2153	557.5	553.2	1114.	1106.
0.2716	555.8	552.2	1109.	1104.
0.3252	548.3	541.2	1092.	1085.
0.4516	528.7	522.6	1047.	1046.
0.5575	500.3	495.8	986.0	984.5
0.7113	462.1	458.9	903.5	902.7
0.8200	436.8	435.7	848.9	845.4
0.9072	420.0	421.1	811.8	815.8
1.	387.9	387.9	745.0	745.0

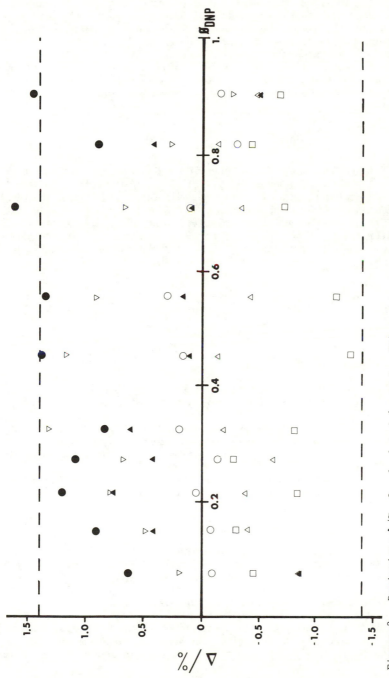

Figure 3. Deviations Δ/% of calculated (equation 22) partition coefficients relative to experimental values for the solutes: n-pentane (closed circles), n-hexane (open circles), n-heptane (open triangles), n-octane (open squares), cyclohexane (open dels), and methylcyclohexane (closed triangles) with squalane + dinonyl phthalate solvents at 30 C. Horizontal dashed lines indicate estimated (47) limits of experimental error.

coefficients of Table II (n-alkane solutes) and equation 22. The relative deviations are then shown as dels (cyclohexane) and filled triangles (methylcyclohexane) in Figure 3. The worst of these are: 1.32% for cyclohexane at $\phi_{DNP} = 0.3252$; and -0.87% for methylcyclohexane at $\phi_{DNP} = 0.0751$.

Having at this point established at least the internal consistency of $\bar{\gamma}_M$ as a function of ϕ_{DNP} for the non-interactive solutes, the data were next used to approximate the excess free energy of mixing of SQ with DNP, as well as the (finite-concentration) activity coefficients $\gamma_{i(M)}$ of each of the solvent components in the stationary phase. In doing so, the volume fractions ϕ_{DNP} were first converted to mole fractions x_{DNP} via the relation:

$$x_{DNP} = \frac{r\ \phi_{DNP}}{1 + \phi_{DNP}\ (r - 1)}$$

where r is the ratio of the molar volumes of SQ (526.69 $cm^3\ mol^{-1}$) and DNP (434.34 $cm^3\ mol^{-1}$). The averaged $\bar{\gamma}_M$ of Table II were then used with equation 21 to calculate g_M^E, the results being smoothed by fitting of the two-term Margules expression, equation 24b, over the regions $x_{DNP} = 0$ to 0.3 and 0.7 to 1.0. This yielded values for the coefficients A and B of 0.3484_4 and 0.4774_1, respectively. The regression of g_M^E as a function of x_{DNP} is shown in Figure 4, and is slightly asymmetric toward the DNP ordinate. Finally, $\gamma_{SQ(M)}$ and $\gamma_{DNP(M)}$ were calculated from the relations:

$$\ln\ \gamma_{SQ(M)} = \left[A + 2(B - A)x_{SQ} \right] x_{DNP}^2 \qquad (25a)$$

$$\ln\ \gamma_{DNP(M)} = \left[B + 2(A - B)x_{DNP} \right] x_{SQ}^2 \qquad (25b)$$

Plots of $\ln\ \gamma_{SQ(M)}/(1 - x_{SQ})^2$ against x_{SQ} and of $\ln\ \gamma_{DNP(M)}/(1 - x_{DNP})^2$ vs. x_{DNP} returned the above-noted values of A and B with correlation coefficients of 0.9_78 and 0.9_96.

The $\gamma_{i(M)}$ are provided below in Table IV; the infinite-dilution activity coefficients derived either by extrapolation of these or by direct calculation (A = $\ln\ \gamma_{SQ(M)}^{\infty}$; B = $\ln\ \gamma_{DNP(M)}^{\infty}$) were thereby found to be 1.416_9 for SQ in DNP, and 1.611_9 for DNP in SQ at 30°C.

It is worth reiterating at this point that these data were determined solely from the infinite-dilution partition coefficients of the (non-interactive) n-alkane probe solutes.

Interactive-Solute Solution with Interactive Solvents

Generally speaking, description of the retentions of solutes that interact stoichiometrically with either (or both) of the solvents, concomitant with solvent-component interaction, would seem to be quite complex. For example, in the instance that B and C form discrete adducts BC in solution (giving rise as a result to negative deviations from Raoult's law), the solute would in effect be subject to partitioning between the gas phase and a ternary-solvent stationary phase. Further, while it might be possible to assess separately the complexation of A with C as well as with B (for example, by determining K_{AB} and K_{AC} in separate series of experiments with an inert

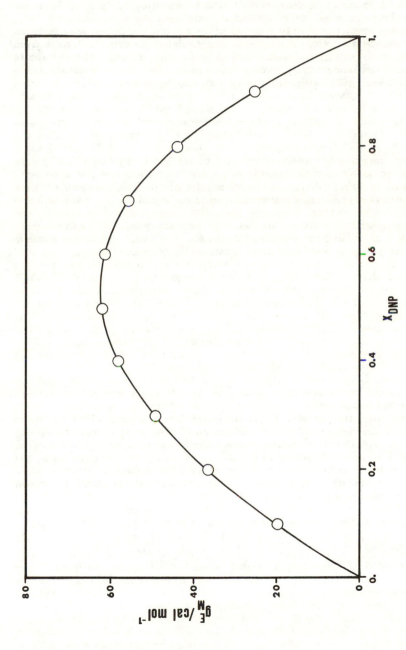

Figure 4. Plot of the excess free energy of mixing g_M^E for squalane + dinonyl phthalate at 30°C, calculated from the two-term Margules expression, equation 24b, with $A = 0.3484_4$ and $B = 0.4774_1$.

diluent phase D, i.e., equations 14 and 15), it would be conjecture to suggest at this time (and with no supportive evidence as yet available) that any excess over the $K_{R\,(M)}^{o}$ calculated for A with B + C (i.e., in excess of that predicted by the relation resulting from substitution of equation 18 into equation 22) would be due to some or other expression involving $K_{A(BC)}$.

In contrast, the situation resulting from solute solution with solvents that interact in a negative sense, that is, give rise to positive deviations from Raoult's law, is straightforward. Supposing for the moment that the solute complexes stoichiometrically with solvent-component C, the notional partition coefficient of A with C, $K_{R(C)}^{o}$, will still be a constant (cf. equation 15c). That is, equation 22 is expected still to be valid. The same reasoning applies as well if the solute interacts both with B and C, viz., equation 18.

The partition-coefficient data for benzene and toluene solutes with squalane + dinonyl phthalate solvents provide an opportunity to test this, the putative K_{AC} with DNP being reported (39) to be 0.276 ± 0.026 and 0.261 ± 0.030 dm^3 mol^{-1}, respectively, at 30°C. [A corroborative value of 0.22 ± 0.10 dm^3 mol^{-1} has also been obtained by NMR for the former (49).] Accordingly, we compare in Table V below the experimental partition coefficients of these solutes with the $K_{R\,(M)}^{o}$ calculated with equation 22 and the $\bar{\gamma}_M$ given in Table II.

The relative deviations for benzene indicate quite remarkable agreement: the largest discrepancy is -0.95% at ϕ_{DNP} = 0.4516, while the average of the deviations is -0.09$_8$%, i.e., equation 22 describes the experimental partition coefficients of this supposed complex-forming solute on average to better than 0.1%. Toluene exhibits an average positive residual of 0.97% (roughly that of cyclohexane); even so, this also lies well within the limits of experiment of ± 1.4%. Moreover, of the total of 20 data points for these two interactive solutes, only 1 (that for toluene at ϕ_{DNP} = 0.2153, 1.91%) falls outside the level taken for random error, and then only marginally so. In fact, of the 80 points considered altogether for SQ + DNP (four n-alkane, two alicyclic hydrocarbon, and two aromatic hydrocarbon solutes), the average residual of the deviations overall amounts only to 0.22%.

In terms of the isotherm formalism, the situation with regard to inert as well as interactive solutes with interactive solvents amounts to that with inert and interactive solutes with non-interactive solvents. That is, since equation 22 appears to provide a description of the partition coefficients of systems of the latter to within experimental error, empirical fitting of equation 13 would seem to be superfluous. However, it may not always be the case that solute and/or solvent positive or negative deviations from Raoult's law can be defined separately. This is particularly true of liquid chromatography, wherein the mechanisms of solute retention are at best only poorly defined. Thus, and for the purpose of achieving analytical separations, the isotherm relation may in fact prove to be advantageous insofar as it is a well-defined continuous function and, hence, can be incorporated immediately into the Laub-Purnell window-diagram scheme of optimization. We consider the matter in further detail in the Section that follows.

Table IV. Finite-Concentration Activity Coefficients $\gamma_i(M)$ for Blends (M) of Squalane (SQ) with Dinonyl Phthalate (DNP) at 30°C, Calculated from Mean Solvent Activity Coefficient Data of Table II and Equations 25

x_{DNP}	$\gamma_{SQ(M)}$	$\gamma_{DNP(M)}$
0.	1.	1.611_9
0.0896	1.004_7	1.457_2
0.1743	1.017_2	1.342_9
0.2497	1.034_4	1.261_8
0.3114	1.052_3	1.207_2
0.3688	1.072_0	1.164_5
0.4996	1.126_6	1.091_2
0.6044	1.178_9	1.051_6
0.7492	1.261_0	1.018_0
0.8467	1.320_7	1.006_1
0.9222	1.368_1	1.001_5
1.	1.416_9	1.

Table V. Comparison of Experimental Liquid–Gas Partition Coefficients of Benzene and Toluene Solutes with $K^o_{R(M)}$ Calculated from Mean Solvent Activity Coefficient Data of Table II and Equation 22 for Blends of SQ + DNP at 30°C

	$K^o_R (M)$					
	Benzene			Toluene		
ϕ_{DNP}	Calc.	Exptl.	Δ/%	Calc.	Exptl.	Δ/%
0.	434.3	434.3	0.	1428.	1428.	0.
0.0751	455.6	455.4	−0.04	1489.	1504.	1.01
0.1483	479.5	481.4	0.40	1558.	1574.	1.03
0.2153	502.0	505.7	0.74	1623.	1654.	1.91
0.2716	524.5	525.0	0.10	1689.	1708.	1.13
0.3252	540.9	539.4	−0.28	1736.	1760.	1.38
0.4516	578.8	573.3	−0.95	1843.	1844.	0.05
0.5575	597.2	597.1	−0.02	1891.	1921.	1.59
0.7113	625.5	621.9	−0.58	1965.	1976.	0.56
0.8200	646.6	647.7	0.17	2022.	2040.	0.89
0.9072	668.4	664.9	−0.52	2082.	2085.	0.14
1.	667.1	667.1	0.	2070.	2070.	0.

APPLICATION OF EQUATION 10 TO LIQUID CHROMATOGRAPHY

In the instance of chromatography generally, the chemical potentials (μ) of the solute (species A) in the stationary (S) and mobile (M) phases must be equal in order to fulfill the condition of thermodynamic equilibrium:

$$\mu_A^M = \mu_A^S \tag{26}$$

Or, upon identifying the standard-state chemical potentials pertinent to liquid chromatography, and replacing activities with molarity-based concentrations and activity coefficients, equation 26 becomes (50,51):

$$c_A^M \, \gamma_A^{M,\infty} = c_A^S \, \gamma_A^{S,\infty} \tag{27}$$

where the superscript ∞ indicates infinite dilution. The solute partition coefficient K_R^o is defined as c_A^S/c_A^M, whereupon substitution of equation 27 for the concentration terms yields:

$$K_R^o = \gamma_A^{M,\infty} \Big/ \gamma_A^{S,\infty} \tag{28}$$

It has now been confirmed in very many studies both of "normal" as well as "reverse-phase" isochratic liquid chromatography that, for a given binary carrier B + C, the activity coefficient of the solute in the stationary phase is approximately independent of compositional variation of the mobile phase. For example, in the particular instance of silica LC with, say, ethyl acetate + hexane, the adsorbent surface sorbs a saturation layer of pure ethyl acetate at concentrations of it of as low as 3% v/v in the mobile phase. Moreover, the sorbed surface layer appears to be insensitive to any further addition of ethyl acetate (52,53). In "reverse-phase" LC, there is in contrast some evidence to suggest that the physical state, e.g., of C_{18} chains is affected by the composition of the carrier. Thus, such chains may well "flatten" to some extent at high concentrations of, for example, water in admixture with methanol (54,55). Even so, whatever change there might then be of the solute activity coefficient in the stationary phase appears to be small in comparison with that in the mobile phase.

Accordingly, for nonionic solutes and binary mobile phases B + C, equation 28 is rewritten as:

$$\frac{1}{K_{R(M)}^o} = \frac{\gamma_{A(M)}^{S,\infty}}{\gamma_{A(M)}^{M,\infty}} \quad \propto \quad \phi_C \tag{29}$$

Thus, inverse retentions are predicted to vary linearly with ϕ_C, the volume fraction of mobile-phase component C in mixtures M of B + C, so long as the solute activity coefficient in the stationary phase remains approximately constant.

Equation 29, first derived and discussed by Scott and Kucera (52,53), is clearly the LC analog of the GC relation, equation 12. And, as with the latter, while it is found to hold in many instances, it fails badly in very many others. Madden, McCann, Purnell, and Wellington (56-58) were the first to address this problem from the standpoint of the Langmuir model. Essentially, they employed a one-term version (n = 1) of equation 13, cast as the inverse in accordance with equation 29. Subsequently, Hsu, Laub, and Madden (59,60) extended this to the full summation form, viz.,

$$\frac{1}{k'_{A(M)}} = \phi_C \left\{ \frac{1}{k'_{A(C)}} + \sum_{i=1}^{n} \left[\frac{b_i \, \phi_B}{1 + b'_i \, \phi_B} \right] \right\} + \frac{\phi_B}{k'_{A(B)}} \tag{30}$$

where $k'_{A(i)}$, the solute capacity factor with mobile phase i, is equal to K^o_R/β, and where β is the mobile-phase/stationary-phase volume ratio. They then showed that equation 30, with n < 2, provides fits of all known curve shapes of inverse retentions against mobile-phase composition to date encountered in LC, including those that are fully parabolic:

$$\frac{1}{k'_{A(M)}} = \phi_C \left[\frac{1}{k'_{A(C)}} + \frac{b_1 \, \phi_B}{1 + b'_1 \, \phi_B} \right.$$

$$\left. + \frac{b_2 \, \phi_B}{1 + b'_2 \, \phi_B} \right] + \frac{\phi_B}{k'_{A(B)}} \tag{31}$$

Further, the ability to fit retentions as a function of this or that system parameter (cast as hypothetical "ϕ") provides the means of computer optimization, via window diagrams, not only of the mobile phase but, in addition, the column temperature, gradient programs, and so forth (61,62).

However, it was found in the latter work that the roots b_i, b'_i for several systems were complex; equation 31 was therefore rearranged to the form:

$$\frac{1}{k'_{A(M)}} = \phi_C \left[\frac{1}{k'_{A(C)}} + \frac{c_1 \, \phi_B + c_2 \, \phi_B^2}{1 + c'_1 \, \phi_B + c'_2 \, \phi_B^2} \right] + \frac{\phi_B}{k'_{A(B)}} \tag{32}$$

where the following substitutions have been made:

$$c_1 = (b_1 + b_2) \tag{33a}$$

$$c'_1 = (b'_1 + b'_2) \tag{33b}$$

$$c_2 = (b_1 b'_2 + b_2 b'_1) \tag{33c}$$

$$c'_2 = (b'_1 b'_2) \tag{33d}$$

It is therefore not surprising that equation 31 fits a variety of curve shapes since, with the Langmuir term expanded to n = 2 (i.e., equation 32), the relation in fact contains the ratio of two quadratics. Nevertheless, and particularly in view of the successful quantitative interpretation of the Langmuir forms of isotherm equations in the instances of the BET model (equations 7, et seq.) and retentions with blended stationary phases in gas chromatography (equations 14, et seq.), the LC relations cannot be dismissed as entirely empirical since, in any event, although such a connection has yet to be established, whatever interpretations are placed on the fitting constants must presumably involve at least the solute activity coefficients in the mobile and stationary phases (see below) and, most likely, the (finite-concentration) activity coefficients pertinent to the mobile- and stationary-phase components as well.

APPLICATION OF EQUATION 10 TO ION EXCHANGE CHROMATOGRAPHY

Of the various forms of liquid chromatography, that involving ion exchange would seem to be most amenable to a precise physicochemical description since the mobile-phase/stationary-phase interaction amounts simply to displacement of one (matrix-bound) ion by another. However, the effects introduced by charged species render such systems even more intractable than is the case with non-electrolyte solutions. Nevertheless, we show in what follows that a rigorous expression can be written for the relevant equilibrium constant, and that this can be related directly to chromatographic retentions.

Consider an anionic solute A of valency y, A^{y-}, in competition with an anionic mobile-phase additive C of valency x, C^{x-}, for sites of exchange. Following the nomenclature of Haddad and Cowie (63) we write:

$$x\ A_M^{y-} + y\ C_S^{x-} \rightleftharpoons x\ A_S^{y-} + y\ C_M^{x-}$$

where x and y are stoichiometric coefficients, and where the subscripts S and M refer to stationary-matrix and mobile phases, respectively. The thermodynamic equilibrium constant K in terms of activities a is then given by:

$$K = \frac{\left(a_{A_S^{y-}}\right)^x \left(a_{C_M^{x-}}\right)^y}{\left(a_{A_M^{y-}}\right)^x \left(a_{C_S^{x-}}\right)^y} \tag{34}$$

We now replace the activities with the products of concentrations and activity coefficients (cf. equation 27), viz.,

$$K = \frac{\left(c_{A^{y-}}^S\ \gamma_{A^{y-}}^{S,\infty}\right)^x \left(c_{C^{x-}}^M\ \gamma_{C^{x-}}^M\right)^y}{\left(c_{A^{y-}}^M\ \gamma_{A^{y-}}^{M,\infty}\right)^x \left(c_{C^{x-}}^S\ \gamma_{C^{x-}}^S\right)^y} \tag{35}$$

where a superscript infinity is appended only to the activity coefficients of the solute, that is, allowance is made for finite concentration of the ionic mobile-phase additive.

Rearrangement of equation 35 next provides:

$$\left[\frac{C_{A^{y-}}^{S}}{C_{A^{y-}}^{M}}\right]^{x} = K \left[\frac{C_{C^{x-}}^{S}}{C_{C^{x-}}^{M}}\right]^{y} \left[\frac{\gamma_{C^{x-}}^{S}}{\gamma_{C^{x-}}^{M}}\right]^{y} \left[\frac{\gamma_{A^{y-}}^{M,\infty}}{\gamma_{A^{y-}}^{S,\infty}}\right]^{x} \tag{36}$$

Then, taking the xth root of both sides,

$$K_{R}^{o} = \left[\frac{C_{A^{y-}}^{S}}{C_{A^{y-}}^{M}}\right] = K^{\frac{1}{x}} \left\{\left[\frac{C_{C^{x-}}^{S}}{C_{C^{x-}}^{M}}\right]\left[\frac{\gamma_{C^{x-}}^{S}}{\gamma_{C^{x-}}^{M}}\right]\right\}^{\frac{y}{x}} \left[\frac{\gamma_{A^{y-}}^{M,\infty}}{\gamma_{A^{y-}}^{S,\infty}}\right] \tag{37}$$

Inverting equation 37 in order to cast it in the form of equation 29, followed by replacement of the partition coefficient with the capacity factor, thereby yields the expression:

$$\frac{1}{k'_{A(M)}} = (\text{const.})\left(C_{C^{x-}}^{M}\right)^{\frac{y}{x}} \tag{38}$$

Equation 38 can be expected to hold in instances where the mobile-phase additive B is present in large excess over A such that the solute and additive activity coefficients as well as $C_{C^{x-}}^{S}$ remain invariant with $C_{C^{x-}}^{M}$.

The function can of course also be written in terms of logarithms:

$$\log k'_{A(M)} = (\text{const.'}) - \frac{y}{x} \log\left(C_{C^{x-}}^{M}\right) \tag{39}$$

the advantage being that a linear relation is obtained. However, plotting the retention data in this manner amounts essentially to graphical comparison of logs which, as we show below, masks deviations in the manner of regression of $1/(k'_{A(M)})$ against mobile-phase composition.

We now seek to cast equation 38 in the form of equation 30. To begin, we let mobile-phase component C be the ionic additive as above; component B then represents the mobile phase without additive. Accordingly, in the case of ion exchange, $\phi_{B}/k'_{A(B)}$ must be zero since solute ions would be retained indefinitely in the absence of finite-concentration additive species competing to displace them from matrix sites of exchange. Next, we set b'_{1} equal to zero and, in considering the simplest case, let n = 1, i.e., a single-term expansion of the summation. Manipulation of the relations, and recollection that $\phi_{B} = (1 - \phi_{C})$, then yields a generalized expression for b_{1}:

$$b_1 = -\text{const.} \left[\frac{1 - (\phi_C)^{\frac{y}{x}-1}}{1 - \phi_C} \right] \tag{40}$$

Or, in the absence of excess volumes of mixing, recognizing that $\phi = C_i \bar{V}_i$ ($\bar{V}_i$ being a molar volume), and replacing ϕ_C with $C^M_{C^{x-}}$, equation 40 becomes:

$$\frac{1}{k'_{A(M)}} = C^M_{C^{x-}} \left\{ \text{const.}' \right.$$

$$\left. - \text{const.}' \left[\frac{1 - \left(C^M_{C^{x-}}\right)^{\frac{y}{x}-1}}{1 - C^M_{C^{x-}}} \right] \left(1 - C^M_{C^{x-}}\right) \right\} \tag{41}$$

Equation 41 thereby provides the means of representing solute inverse retentions as a function of the concentration of mobile-phase additive irrespective of the valency of either. For example, in those instances where y = x, the relation reduces to:

$$\frac{1}{k'_{A(M)}} = \text{const.}' \left(C^M_{C^{x-}}\right) \tag{42a}$$

whereas, for y = 2x,

$$\frac{1}{k'_{A(M)}} = \text{const.}' \left(C^M_{C^{x-}}\right)^2 \tag{42b}$$

Alternatively, for y = ½x,

$$\frac{1}{k'_{A(M)}} = \text{const.}' \left(C^M_{B^{x-}}\right)^{½} \tag{42c}$$

In the trivial case y = 0 (x finite), the relation also predicts that inverse retentions will be invariant with mobile-phase additive concentration:

$$\frac{1}{k'_{A(M)}} = \text{const.}' \tag{42d}$$

It must again be emphasized, however, that all of these expressions will be valid only in the limit that the solute and additive activity coefficients do not vary with additive concentration.

Test of Equation 39

The ion-exchange data reported by Haddad and Cowie for a series of anions with phthalate/biphthalate additive at pH 5.3 (63) comprise an opportunity for testing equation 39 (as well as equations 41, 42a, and 42c; see below). Accordingly, we provide in Table VI log $k'_{A(M)}$ as a function of log C^M_C for the systems they considered, as well as the slope m, the y-intercept b, and the absolute value of the linear least-squares correlation coefficient r, the data being derived from Figure 1 of their work.

Table VI. Log $k'_{A(M)}$, Slope m, y-intercept b, and Linear Least-Squares Regression Correlation Coefficient r for Indicated Anions as a Function of Log (Phthalic Acid Mobile-Phase Additive Concentration) at pH 5.3. Data Interpolated from Figure 1 of Ref. 63

Anion	$\log k'_{A(M)}$ at $-\log(H_2P) =$					$-m$	$-b$	r
	2.70	2.40	2.22	2.10	2.00			
Cl^-	0.00952_3	-0.137_0	-0.257_1	-0.350_5	-0.371_4	0.571_9	1.527_5	0.9994_5
Br^-	0.238_1	0.0438_0	-0.0762_0	-0.139_1	-0.198_1	0.625_5	1.455_0	0.9999_3
$H_2PO_4^-$	0.430_8	0.203_8	0.120_0	-0.0114_3	-0.0591_0	0.664_8	1.386_9	0.9994_7
SO_4^{2-}	0.761_9	0.426_7	0.209_5	0.102_9	-0.02667_7	1.119_8	2.262_8	0.9999_5
$S_2O_3^{2-}$	0.916_2	0.590_5	0.373_3	-0.261_0	0.146_7	1.101_4	2.058_1	0.9999_7
I^-	0.657_1	0.472_4	0.356_2	0.291_4	0.238_1	0.603_4	0.975_2	0.9999_5
$C_2O_4^{2-}$	0.942_9	0.655_2	0.476_2	0.382_9	0.276_2	0.946_8	1.615_7	0.9999_6
"System"[a]	1.000_0	0.767_6	0.630_5	0.537_1	0.478_1	0.754_1	1.039_6	0.9999_5

[a] The magnitude of the "system" peak was said to vary as a function of the type and concentration of sample ion; the injection volume, the pH of the analyte and eluent solutions, and so forth.

We note, first, that the correlation coefficients, for what they are worth (64), for the log-plots of Cl^- and $H_2PO_4^-$ are indicative of some scatter in the data; even so, they (as well as the remaining solutes) appear to conform to linear regression, as expected in light of the log-form of equation 39.

Secondly, Haddad and Cowie reported that phthalic acid exists in solution as 47.6% HP^-, 52.2% P^{2-}, and 0.2% H_2P at the temperature of their experiment and so, plots of $log(k'_{A(M)})$ against $log\ C_C^M$ (i.e., equation 39) for the univalent solutes would be expected to have slopes either of -1.0, of -0.66, or of -0.50, depending upon whether HP^-, $(HP^- + P^{2-})$, or P^{2-} governed the elution behavior, that is, controlled the solute retentions. The slopes actually observed for Cl^- (0.57), Br^- (0.63), $H_2PO_4^-$ (0.67), and I^- (0.60) are in fact only approximately consistent but, on average, approach -0.66. Thus, for these solutes at least, one interpretation of the log data might be that the blend of species $(HP^- + P^{2-})$ predominates in controlling the retentions. However, this does not appear to be the case in the instances of the divalent solutes: the theoretical slopes of equation 39 with HP^-, $(HP^- + P^{2-})$, and P^{2-} are -2.00, -1.33, and -1.00, respectively, whereas those found were -1.13 (SO_4^{2-}), -1.10 ($S_2O_3^{2-}$), and -0.98 ($C_2O_4^{2-}$) which fall closest to that predicted for P^{2-} mobile-phase additive acting alone.

Test of Equations 41 and 42

The alternative view of these retention data, namely, the isotherm relation, equation 41, is shown tested in Figure 5, where the experimental inverse capacity factors are plotted against C_C^M. (The abscissa could be converted to the concentration either of HP^- or of P^{2-} with the appropriate scaling factor although, since each is directly proportional to the concentration of H_2P, there would be little advantage, indeed some inconvenience, in doing so.) We see, first, that virtually without exception all of the points for the divalent species (as well as those for the "system") fall on straight lines, that is, correspond to equation 42a for which y = x. Thus, the divalent mobile phase additive P^{2-} appears to control entirely the retentions of these solutes.

Figure 5 also leaves little doubt that the data for the univalent species describe curves that are concave to the abscissa. In addition, the points for Cl^- at 4 mM and at 8 mM H_2P appear to be slightly discrepant, while the datum for $H_2PO_4^-$ at 6 mM is clearly very much so. Nevertheless, we assume at this point that, as with the divalent species, P^{2-} controls the retentions of these solutes, i.e., that y = ½x and that equation 42c therefore applies. Thus, plots of $(k'_{A(M)})^{-1}$ against $(H_2P)^{1/2}$ should be linear for these monovalent ions. In order to test this, the inverse capacity-factor data derived from least-squares treatment of the experimental values of Table VI (which in any event must be linear when regressed in log form against log mobile-phase additive) are shown plotted in Figure 6 (open circles) against $(H_2P)^{1/2}$: the data obviously describe straight lines. Next, the actual experimental data were plotted as solid circles. These, too, appear to regress linearly against $(H_2P)^{1/2}$ as predicted. Moreover, the two points for Cl^- as well as that for $H_2PO_4^-$ which were said earlier to be potentially discrepant are thereby brought out clearly as being so, the latter to about the same extent as the former two.

Also shown for comparison as triangles and dashed curves in Figure 6 are the experimental data for SO_4^{2-}, $S_2O_3^{2-}$, and $C_2O_4^{2-}$ which, according to equations 42, should regress against $(H_2P)^{1/2}$ convex to the abscissa. That this is the case provides further evidence that the retentions both of the divalent (y = x) and the univalent (y = ½x) solute anions are controlled solely by P^{2-}, and that equation 42a describes the former while equation 42c applies to the latter.

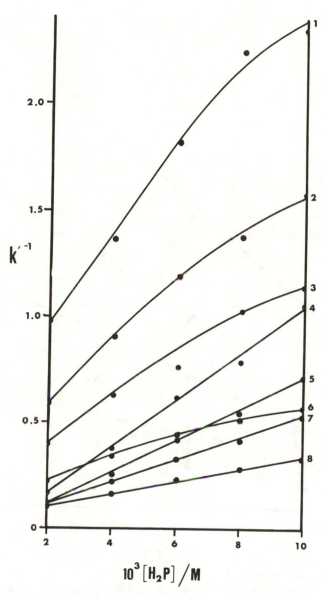

Figure 5. Plots of experimental $(k'_{A\,(M)})^{-1}$ (cf. Table VI) against concentration (mM) of H_2P mobile-phase additive for the anion solutes: Cl^- (1), Br^- (2), $H_2PO_4^-$ (3), SO_4^{2-} (4), $S_2O_3^{2-}$ (5), I^- (6), $C_2O_4^{2-}$ (7), and the "system" peak (8).

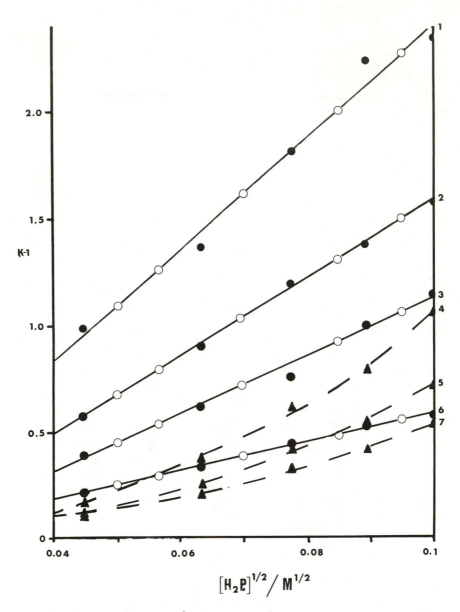

Figure 6. Plots of $(k'_{A(M)})^{-1}$ against $(H_2P)^{1/2}$, equation 42c. Open circles, lines: inverse capacity-factor data for the monovalent anions derived from least-squares regression of experimental values of Table VI and equation 39; filled circles: observed inverse capacity factors. Triangles, dashed curves: observed inverse capacity factors of the divalent anions. Solutes numbered as in Figure 5.

The generalized Langmuir–isotherm relation in the form of equation 41 thereby appears to encompass ion exchange chromatography in addition to GC and LC; most important, and as was the case for GC, each of the fitting parameters can be interpreted quantitatively. In this instance, they pertain to the solute and additive concentrations and activity coefficients in the mobile and stationary phases (equation 37).

SUMMARY AND CONCLUSIONS

By definition, any model, however closely it approximates reality, remains only an analogy. Thus, for example, modeling infinite–dilution activity coefficients by assuming that atoms (or segments of molecules) occupy positions of a three-dimensional lattice and that interactions occur only at cross-points, offers at best only a crude picture of the continuum of intermolecular interactions in the liquid state. Lattice models have therefore been roundly criticized on occasion; even so, the results thereby predicted, at least for systems for which the equations are tractable (i.e., those comprised of n–alkanes; cf. refs. 40,65) far exceed those arising, e.g., from solubility-parameter theory.

Much the same can also be said of the use made in this work of the Langmuir model of sorption. The modified form of the relation, equation 7, cast so as to correct the worst of the approximations made in deriving equation 5, has been shown to encompass the BET formalism and, thereby, isotherm Types 0 through V. Further, when expanded as a summation, equation 10, (multi-stepped) Types VI were accounted for as well. The form of the latter was then shown to be applicable to chromatography when put in terms relevant to the partitioning process, equations 16, 19, and 22 (GC); 30-32 (LC); and 38, 40, and 41 (ion exchange). In addition, the fitting parameters were demonstrated in the first and third of these cases to be related explicitly to the solute and solvent activity and partition coefficients. (This portends that, at the very least, the forecasting of retentions will henceforth require only a minimal body of predetermined data, followed by separation of the solutes of interest by application of the straightforward and quantitative technique of window–diagram optimization.) On the other hand, and despite successful exploitation of the isotherm equations developed in this work and elsewhere we can, at this point, state with confidence only that the form of the relations appears to be applicable to a range of physicochemical phenomena pertaining to sorption; that the relations themselves describe the results arising from experiment; and that the parameters comprising the powers and coefficients can be interpreted exactly in at least some instances. The matter thus appears to warrant further and comprehensive study.

Acknowledgments

Financial support for this work, provided in part by the Department of Energy, Office of Basic Energy Sciences (liquid chromatography) and by the National Science Foundation (gas chromatography), is gratefully acknowledged.

Literature Cited

1. Giddings, J. C. "Dynamics of Chromatography, Part 1, Principles and Theory"; Dekker: New York, 1965.
2. Supina, W. R. In "Modern Practice of Gas Chromatography"; Grob, R. L., Ed.; Wiley-Interscience: New York, 1977; p. 137.
3. Purnell, J. H. "Gas Chromatography"; Wiley: London, 1962.
4. Laub, R. J.; Pecsok, R. L. "Physicochemical Applications of Gas Chromatography"; Wiley-Interscience: New York, 1978.
5. Conder, J. R.; Young, C. L. "Physicochemical Measurement by Gas Chromatography"; Wiley: Chichester, England, 1979.
6. Locke, D. C. In "Advances in Chromatography"; Giddings, J. C., Keller, R. A., Eds.; Marcel Dekker: New York, 1969; Vol. 8, pp. 47-89.
7. Locke, D. C. In "Advances in Chromatography"; Giddings, J. C., Keller, R. A., Eds.; Marcel Dekker: New York, 1978; Vol. 14, pp. 87-198.
8. Laub, R. J.; Purnell, J. H. J. Chromatogr. 1975, 112, 71-79.
9. Laub, R. J. Am. Lab. 1981, 13(3), 47-58.
10. Laub, R. J. In "Physical Methods of Modern Chemical Analysis"; Kuwana, T., Ed.; Academic Press: New York, 1983; Vol. 3, Ch. 4.
11. Langmuir, I. J. Am. Chem. Soc. 1916, 38, 2221-2295.
12. Brunauer, S.; Emmett, P. H.; Teller, E. J. Am. Chem. Soc. 1938, 60, 309-319.
13. Sheng, P. Liq. Cryst. Ordered Fluids 1984, 4, 889-899.
14. Drew, C. M.; Bens, E. M. In "Gas Chromatography 1968"; Harbourn, C. L. A., Ed.; Institute of Petroleum: London, 1969, 3-18.
15. Soriaga, M. P.; Hubbard, A. T. J. Am. Chem. Soc. 1982, 104, 2735-2742; 3937-3945.
16. Tinkle, M.; Dumesic, J. A. J. Phys. Chem. 1983, 87, 3557-3562.
17. Vint, G. J. S.; Phillips, C. S. G. J. Chromatogr. 1984, 292, 263-271.
18. Laub, R. J. Anal. Chem. 1984, 56, 2110-2115; 2115-2119.
19. Purnell, J. H.; Vargas de Andrade, J. M. J. Am. Chem. Soc. 1974, 97, 3585-3590, 3590-3593.
20. Laub, R. J.; Purnell, J. H. J. Am. Chem. Soc. 1975, 98, 30-35, 35-39.
21. Laub, R. J.; Wellington, C. A. In "Molecular Association"; Foster, R., Ed.; Academic Press: London, 1979; Vol. 2, Ch. 3.
22. Laub, R. J. In "Inorganic Chromatographic Analysis"; Wiley-Interscience: New York, 1985; Ch. 2.
23. Laub, R. J. Adv. Instrum. 1980, 35(2), 11-20.
24. Laub, R. J. Glas. Hem. Drus. Beograd 1983, 48, 377-391.
25. Purnell, J. H. In "Gas Chromatography 1966"; Littlewood, A. B., Ed.; Institute of Petroleum: London, 1967, 3-18.
26. Laub, R. J.; Purnell, J. H.; Summers, D. M. J. Chem. Soc. Faraday Trans. I 1980, 76, 362-367.
27. Hildebrand, G. P.; Reilley, C. N. Anal. Chem. 1964, 36, 47-58.
28. Villalobos, R.; Brace, R. O.; Johns, T. In "Gas Chromatography 1959"; Noebels, H. J., Wall, R. F.; Brenner, F., Eds.; Academic Press: New York, 1961, 39-54.
29. Villalobos, R.; Turner, G. S. In "Gas Chromatography 1963"; Fowler, L., Ed; Academic Press, New York, 1964, 105-118.
30. Deans, D. R. Chromatographia 1968, 1, 18-22.
31. Deans, D. R.; Scott, I. Anal. Chem. 1973, 45, 1137-1141.
32. Purnell, J. H.; Williams, P.S. J. High Resolut. Chromatogr. Chromatogr. Commun. 1983, 6, 569-570.
33. Purnell, J. H.; Williams, P. S. J. Chromatogr. 1985, 321, 249-254.

34. Gil-Av, E.; Herling, J. J. Phys. Chem. **1962**, 66, 1208-1209.
35. Martire, D. E.; Riedl, P. J. Phys. Chem. **1968**, 72, 3478-3488.
36. Laub, R. J.; Purnell, J. H. Anal. Chem. **1976**, 48, 799-803.
37. Ali, S. G. A. H.; Purnell, J. H.; Williams, P. S. J. Chromatogr. **1984**, 302, 119-133.
38. Cadogan, D. F.; Sawyer, D. T. Anal. Chem. **1971**, 43, 941-943.
39. Harbison, M. W. P.; Laub, R. J.; Martire, D. E.; Purnell, J. H.; Williams, P. S. J. Phys. Chem. **1979**, 83, 1262-1268.
40. Laub, R. J.; Martire, D. E.; Purnell, J. H. J. Chem. Soc. Faraday Trans. II **1978**, 74, 213-221.
41. Chien, C.-F.; Laub, R. J. Glas. Hem. Drus. Beograd **1983**, 48, 319-333.
42. Moore, W. J. "Physical Chemistry", Third Ed.; Prentice-Hall: Englewood Cliffs, New Jersey, 1962, p. 149.
43. Vernier, P.; Raimbault, C.; Renon, H. J. Chim. Phys. Biol. **1969**, 66, 960-969.
44. "International Critical Tables of Numerical Data, Physics, Chemistry and Technology", Vol. III; Washburn, E. W., Ed.; McGraw-Hill: New York, 1928, p. 182.
45. Furio, D. L. Ph.D. Thesis, The Ohio State University, 1985. Furio, D. L.; Laub, R. J. Work to be published.
46. Dreisbach, R. R. "Physical Properties of Chemical Compounds", Vol. 1; American Chemical Society: Washington, D.C., 1955, pp. 273, 335.
47. Laub, R. J.; Purnell, J. H.; Williams, P. S.; Harbison, M. W. P.; Martire, D. E. J. Chromatogr. **1978**, 155, 233-240.
48. Ashworth, A. J. J. Chem. Soc., Faraday Trans. 1 **1973**, 69, 459-466.
49. Martire, D. E.; Sheridan, J. P.; King, J. W.; O'Donnell, S. E. J. Am. Chem. Soc. **1976**, 98, 3101-3106.
50. Locke, D. C. J. Chromatogr. Sci. **1974**, 12, 433-437.
51. Locke, D. C.; Martire, D. E. Anal. Chem. **1967**, 39, 921-925; **1971**, 43, 68-73.
52. Scott, R. P. W.; Kucera, P. J. Chromatogr. **1975**, 112, 425-442.
53. Scott, R. P. W.; Kucera, P. J. Chromatogr. **1977**, 142, 213-232.
54. Boehm, R. E.; Martire, D. E. J. Phys. Chem. **1980**, 84, 3620-3630.
55. Martire, D. E.; Boehm, R. E. J. Phys. Chem. **1983**, 87, 1045-1062.
56. McCann, M.; Purnell, J. H.; Wellington, C. A. In "Chromatography, Equilibria, and Kinetics"; Young, D. A., Ed; The Royal Society of Chemistry: London, 1980, 83-91.
57. McCann, M.; Madden, S. J.; Purnell, J. H.; Wellington, C. A. Paper presented at the 184th National Meeting of the American Chemical Society, Kansas City, Missouri, 1982.
58. McCann, M.; Madden, S. J.; Purnell, J. H.; Wellington, C. A. J. Chromatogr. **1984**, 294, 349-356.
59. Hsu, A.-J.; Laub, R. J.; Madden, S. J. J. Liq. Chromatogr. **1984**, 7, 615-637.
60. Laub, R. J.; Madden, S. J. J. Liq. Chromatogr. **1985**, 8, 155-174.
61. Deming, S. N.; Turoff, M. L. H. Anal. Chem. **1978**, 50, 546-548.
62. Laub, R. J. J. Liq. Chromatogr. **1984**, 7, 647-660.
63. Haddad, P. R.; Cowie, C. E. J. Chromatogr. **1984**, 303, 321-330.
64. Tiley, P. F. Chem. Brit. **1985**, 21, 162-163.
65. Laub, R. J.; Martire, D. E.; Purnell, J. H. J. Chem. Soc. Faraday Trans. I **1977**, 73, 1685-1690.

RECEIVED June 5, 1985

2

Characterization of Bonded High-Performance Liquid Chromatographic Stationary Phases

S. D. Fazio, J. B. Crowther[1], and R. A. Hartwick

Department of Chemistry, Rutgers, The State University of New Jersey, New Brunswick, NJ 08903

Capillary gas chromatographic methods are described for the analysis of chemically bonded reversed-phase HPLC stationary phase materials. Procedures for acid digestion of the packing material producing stable and volatile derivatives easily analyzed by gas chromatography are compared for precision and accuracy. The methods were used to analyze a variety of commercial phases, possessing widespread monofunctional silane chemistries and bonding consistencies.

The chromatographic properties of chemically bonded packing materials for HPLC can be profoundly influenced by both the physical and chemical properties of the silica support (1-4). The pore size, pore size distribution, surface area and trace metal content (5-7) will influence the surface character of the base silica material. In addition, the concentration of residual silanols (8-13) and the exact chemical composition and environment of the bonded ligand groups (14-22) will effect changes in the selectivity, even between batches of identically bonded materials (23). In recent work in our laboratory, we have been investigating the behavior of blended stationary phases in HPLC (24,25), whereby two or more different ligands are chemically bonded to the same support material. For this type of work, as well as for comparisons to and between commercially available bonded phases, it was necessary to know the exact chemical composition of the final bonded support. Not surprisingly, many manufacturers either will not divulge, or do not know the exact composition of their commercially available materials. Therefore, it was desired that simple, rapid methods be developed to analyze the bonded materials on a routine bases.

[1]Current address: College of Veterinary Medicine, Cornell University, Ithaca, NY 14853

The chemical analysis of bonded siliceous packing materials for HPLC has been accomplished by a variety of techniques. Simple tests based upon chromatographic behavior of test compounds (26-28) or elemental analyses (29) are essential in following the progress of a particular reaction or for quality control of batches, but yield little detailed structural information concerning the stationary phase composition. IR spectroscopy (30-34), thermal methods (21,35), photoacoustic spectroscopy (36), magic angle spinning and FT-NMR (37,38), direct measurements of exchange capacities (39), alkaline hydrolysis followed by derivatization and GC (40), pyrolysis gas chromatography (41), and most recently ESCA (43) have all been developed, and can be very powerful for the particular information desired.

Gas chromatographic methods seemed the most amenable to routine use and required instrumentation which would be widely available in the average laboratory. The GC-pyrolysis methods were not suitable for our purposes since, with multifunctional phases, it was necessary to determine the relative ratios of the various bonded ligands and capping agents present on the surface. The pyrograms would be too complex to analyze under these conditions without extensive use of mass spectrometry. Likewise, the dissolution of silica gel under alkaline conditions (40), followed by GC analysis, yielded somewhat complex samples which were difficult to interpret even for simple phases consisting of only one ligand type.

Recent work in our laboratory (44) produced an acid hydrolysis/GC method based upon controlled mixed-dimer formation. This technique left the original ligand structure intact but in a dimer form. Although suited for the analysis of the "mixed-mode" or "mixed-ligand" bonded phases (45,46) for which it was developed, a shortcoming of the method was that only bonded phases known to utilize monofunctional silanes in the bonding procedure could be analyzed.

A simple, rapid method was sought which could determine silane identity and quantity on bonded HPLC packing materials without prior knowledge of the chemistry of the sample. The use of trimethylsilanes as protecting groups in peptide synthesis (47,48) showed that the silicon oxygen bond could be selectively cleaved by aqueous fluorides. Widespread work in the area of semiconductors (49) showed that the silicon surface is etched by hydrogen fluoride/nitric acid solutions. Erard et al (50) employed anhydrous hydrogen fluoride in diethyl ether for the selective cleavage of siloxane bonds on bonded silica gels. There is also extensive literature on the analysis of silicone polymers using Lewis acids such a boron trifluoride, aluminum trifluoride or aqueous hydrofluoric acid (51-56).

Based upon these results and preliminary experiments in our laboratory, methanolic hydrofluoric acid solutions were chosen as a digestion/derivatization reagents for the proposed method. The use of an easily prepared aqueous hydrogen fluoride-methanol solution for selected cleavage of the siloxane linkage was found to produce stable and volatile mono-, di-, and tri-fluoro oganosilane derivatives suitable for direct extraction and analysis by either capillary GC or head space packed column GC. Advantages of the method over previously published techniques are the simplicity of

the proposed method which uses only inexpensive reaction vials and
readily available gas chromatographic instrumentation.
Furthermore, information is generated which can quantitate
multiple ligands and with minor changes in procedure, be applied to
a variety of bonded reversed-phase silica supports.

Acid Hydrolysis of Bonded Phases. One of the problems encountered
with acid hydrolysis of silane-bonded phases is that dimerization
of the silane ligands is favored (42), as illustrated in Reaction 1
of Figure 1 and Table I. Such dimerization can be helpful for the

Table I. Kovats Retention Indices for Symmetric Dimer Silanes

Dimers	Indices
$C_5Si-O-SiC$	1379
$C_8Si-O-SiC_8$	1941
$ClC_3Si-O-SiC_3Cl$	1499
$PhC_3Si-O-SiC_3Ph$	2274

C_5Si = dimethylpentylsilane,
C_8Si = dimethyloctylsilane
ClC_3Si = 3-chloropropyldimethylsilane
PhC_3Si = 3-phenylpropyldimethylsilane
Kovats index for silane dimers used to analyze commercial materials
or used as internal standards. Column: SE-30

analysis of simple monoligand phases, or a source of added
complexity in the case of multiple ligands, e.g. capping reagents
or blended stationary phases. Symmetric dimers are formed when
uncapped monofunctional support materials are subjected to acid
hydrolysis. If the support material is end-capped, typically with
trimethylchlorosilane (TMCS), both symmetric and unsymmetric
dimers form in proportions dependent both upon the kinetics of
hydrolysis, and upon the rate of dimer formation. If multiple
ligands are bonded to the support, numerous unique dimers can be
formed, greatly complicating the analysis.
     Three basic approaches have been developed in this study to
eliminate these complications, each one being suited for a
particular type of stationary phase. First, if only the surface
concentration of the primary ligand is of interest, simple acid
hydrolysis with dimerization is used. This approach works best for
determination of the primary ligand, e.g. C_8, and for the
estimation of impurities present. The formation of TMS-C_8 dimers
will be minimal, provided the relative surface concentration of
capping agent is low.
     The second approach involves determination of the relative
concentrations of primary ligand and capping agent. In this case,
excess large diameter (inexpensive) silica bonded with TMS is
hydrolyzed along with the bonded silica of interest. Under these
conditions, the formation of a trimethylsilyl-ether derivative of
the ligand is favored, due to the locally high concentrations of
TMS. TMS derivatives of the silanes were produced in excess of
96% yield. For studies concerning multiple ligand bonding and

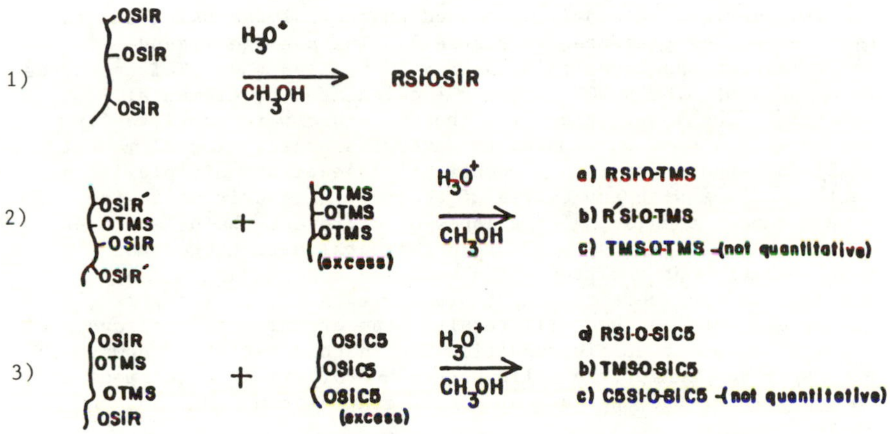

Figure 1. The bonded silica materials are acid cleaved and dimers analyzed using capillary GC according to three procedures, the bonded material can either be 1) acid hydrolyzed and the dimer products extracted and directly analyzed by capillary gas chromatography, or 2) the packing material of interest may be hydrolyzed along with an excess of trimethylsilane (TMS) bonded silica, forming TMS derivatives of the original bonded ligand or ligands, or 3) the packing material is hydrolyzed with excess C_x bonded silica, where C_x is a silane of unique chain length. The C_x derivatives can then be separated and analyzed for the surface concentrations of both capping agent and primary ligands present.

mixed-phases, this approach proved to be simple, producing
excellent qualitative and quantitative results. However, using TMS
as a derivatizing agent obviously precluded obtaining information
concerning the concentrations of TMS capping agent on the silica.

If information concerning the degree of capping was needed. a
third method was developed in which an excess of large particle
silica bonded with a ligand of unique chain length, (for example,
C_5 silane), was added to the packing during the hydrolysis step.
Although this technique is more expensive than using TMS silica,
all ligands of interest and the degree of capping can be determined
in a single GC run by analysis of the C_x derivatives.

Analysis of Reversed-Phase Materials. An example of a simple
dimethyloctylmonochlorosilane bonded phase analyzed using the first
dimer method is presented in Figure 2. The average ligand
concentration was found to be 3.88×10^{-4} moles g^{-1}. This compares
to a value of 5.60×10^{-4} moles g^{-1} calculated by elemental
analysis. The lower values for the GC method were consistently
observed, and were attributed to several sources. The elemental
analysis cannot take into account the presence of multiple ligands
on the surface with any degree of accuracy. Impurities in the
bonded phase, consisting of silanes of different chain length or
other unknown organics, will yield elemental percentages not
representative of the true surface ligand concentration. Also, if
any residual solvents remain absorbed within the silica pores,
higher elemental values will result. The efficiency of cleavage of
the ligands under acidic conditions is another variable; however,
extensive testing of the silica after hydrolysis using elemental
analyses indicated the lack of significant carbon remaining on the
silica.

Analysis of Multifunctional Phases. Figures 3a and 3b show an
application in which the ligand concentrations for a stationary
phase consisting of multiple bonded ligands (in this case a
$C_8/C_3-SO_3^-$ mixed-mode reversed-phase/cation/exchange phase) was
monitored. In the synthesis of these multifunctional phases, it
was necessary to optimizate the sulfonation reaction of the phenyl
ligand with chlorosulfonic acid, and to monitor the extent of
hydrolysis occurring during this step. The degree of sulfonation
was determined by following the decrease in peak area of the
propylphenylsilane-TMS derivative over the course of the
sulfonation. Decrease in dimethyloctylsilane-TMS derivative
reflected loss of ligands due to hydrolysis.

Determination of Capping Agents. An excess of C_5 bonded material
was used to determine the primary bonded ligand plus capping
concentrations on the silica support for several commercial and
laboratory bonded phases (Method 3). A series of bonded phases
were prepared from silane bonding mixtures varying in their TMCS
concentrations from 0% to 89%. The data concerning bonding mixture
percentages versus the actual surface concentrations of the
trimethyl- and dimethyloctyl silanes are presented in Table II.

A commercial C_8 material is included in the list in Table II.
The gas chromatogram of the commercial phase is shown in Figure 4.
It is interesting to note the rather large concentration of TMS

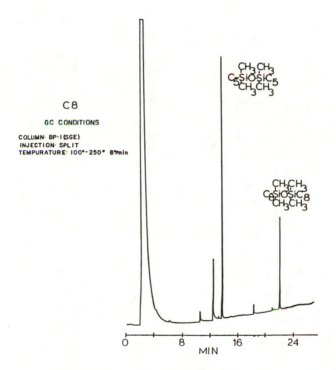

Figure 2. Direct analysis by capillary GC of the acid hydrolysis products of a dimethyloctyl 10um silica phase, using C_5 dimer as an internal standard. The smaller unidentified peaks were from the internal standard and from slight impurities in the hexane extractant. Instrumental and reaction conditions are described in the experimental section.

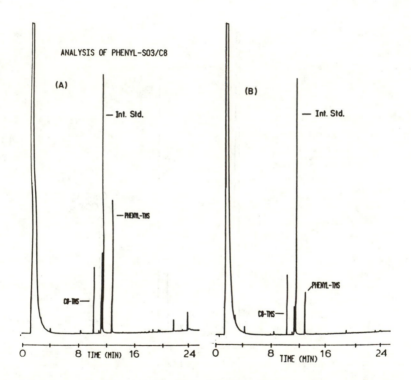

Figure 3. Analysis of chemically bonded mixed phase (a) before
and (b) after sulfonation of the phase with chlorosulfonic acid.
The efficiency of the sulfonation reaction was reflected by the
loss of dimethylpropylphenyl peak area, since the sulfonated
products were not volatile. The hydrolysis of all ligands was
reflected in the loss of dimethyloctyl silane peak area.

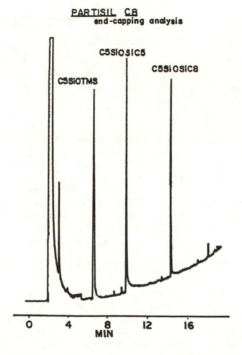

PARTISIL C8
end-capping analysis

C5SIOSIC5

C5SIOSIC8

C5SIOTMS

Figure 4. Analysis of a commercial C_8 material using method 3. An excess of C_5 bonded silica was added to the reversed-phase material during hydrolysis. Under these conditions, the C_5 derivatives of the reversed-phase ligands are preferentially formed, allowing for the direct monitoring of both TMS capping agent and primary ligand identification and quantification. Other conditions as in Figure 2.

relative to dimethlyoctyl groups for this dimethyloctyl phase.
Using direct analytical methods, such as the one presented here,
should yield interesting data concerning various commercial
phases, and the relationships between their particular chemistries
and their chromatographic properties. A more unified understanding
of differences between phases may result as this type of data
becomes available.

Table II. Determination of C_8/TMS Capping Ratios
Using Capillary GC

Bonded Material	%C_8	%TMS	%C Elemental	mole % of Reactants During Bonding	
				C8	TMCS
100:0	100%	0%	9.2%	100%	0%
49:51	3.4%	96.6%	5.4%	49%,	51%
11:89	0%	100%	4.8%	11%	89%
C_8 Commercial	31%	69%	8.5%	--	--

Analysis of a series of test phases and one commercial phase, for
the surface concentrations of dimethyloctyl silane and
trimethylchlorosilane groups was performed using dimethylpentyl
silane as the derivatizing agent. The test silicas were prepared
using the indicated mole ratios of monochlorosilane reagents (last
column), producing the indicated actual coverages (column 2). It
is interesting to note that the commercial material labeled as C_8
in fact has a surface comprised of nearly 50% trimethylsilane
groups. Little information concerning the degree of capping agent
can be discerned from the elemental data alone.

This method was also found to be very useful for checking the
purity of chlorosilane reagents used in the various bonding
procedures. Commercial dimethlyoctyl silanes were generally found
to be fairly pure, while several dimethlyoctadecyl, and
dimethlydodecyl monochlorosilanes had gross impurities of numerous
other chain lengths.

Accuracy and Precision. In general, analysis by any of the three
GC methods described produced ligand concentrations significantly
lower than obtained using elemental analysis, even with carefully
prepared monofunctional phases. As previously discussed, there are
several possible sources for this discrepancy.
     For single batches of bonded phases, relative standard
deviations of between 3-11% were observed. The real usefulness of
the methods was in the qualitative information obtained. Details
concerning exact surface coverages of multiple ligand groups,
relative ratios of capping reagents, and of overall silaca coverage
were obtainable with very simple instrumentation available in most
analytical laboratories. Much of this imformation could otherwise
only be obtained by using other very sophisticated methods, if at
all.

Fluoride Derivatization Technique

Gas Chromatographic Analysis. A simplified reaction scheme showing
the major expected products of fluoride derivatization is shown in
Figure 5. The fluoride derivatives of the silanes are more
volatile than the corresponding chlorides, aiding the gas
chromatographic analysis of the longer chain silane ligands. An
SE-30 stationary phase on a fused silica capillary column was
chosen for the generaly analysis of the fluorosilyl derivatives,
producing excellent results as shown in Figure 6. Analysis of
chain lengths up to about C_{18} for the dimethylmonofluoro derivative
was possible on the SE-30 column. Figure 6 also shows that the
lower limit of chain length was about C_6 trifluoro derivative.
Solutes more volatile than this, such as trimethylfluorosilane,
were analyzed by head space GC.

In order to reduce the time of analysis for routine
reversed-phase analysis, GC conditions were optimized for each
bonded phase chain length. Figures 7a and 7b show chromatograms of
test mixtures for mono-, di- and trifluoro derivatives of octyl and
octadecyl phases, including the internal standards. Analysis times
were reduced to 11 minutes and 18 minutes, respectively, for each
phase type. The response factors for both the octyl and octadecyl
phases were found to be nearly unity for an FID detector.
Conditions were also established for the routine separation of
octyl or octadecyl phases of mono-, di- or trifunctional silanes
using packed columns with an OV-1 phase.

Headspace Analysis of Capping Agents. Separation and analysis of
the TMS "capping" agent often used on stationary phases
necessitated the use of different GC procedures. Headspace
analysis of a separate digest of the bonded phase of interest is
performed, using septa-type Teflon vials. Figure 8 shows a gas
chromatogram of methyl-, ethyl- and n-propyldimethylmono-
fluorosilane derivatives, separated on a Porpak P packed GC column,
with thermal conductivity detection. Good separations were
achieved with total separation times of under 10 minutes.

Hydrofluoric Acid Digestion Kinetics. The kinetics of the
fluoride derivatization reaction for octyl and octadecyl, mono- and
trifluoro-silyl derivatives were studied, as shown in Figure 9,
along with values obtained by elemental analysis for each phase.
Formation of the fluorodimethyloctylsilane species was quantitative
within 2 hrs., while the trifluorooctyl- and octadecylsilanes
required about 4 hrs. for conversion and phase transfer to the
hexane. As compared to the fluorodimethyloctylsilane, nearly
identical kinetics were observed for the fluorodimethyloctadecyl-
silane derivatives. Distribution coefficients for the
monofluoroderivatives and the internal standards, between the
aqueous acid and hexane were measured, and found to exceed 99%.

An apparent recovery problem was consistently observed for the
trifluorooctylsilane and trifluorooctadecylsilanes, as shown in
Figure 9. The yield of the ligand coverage when using the
trifluoro- derivative of either the octyl or octadecyl phase alone

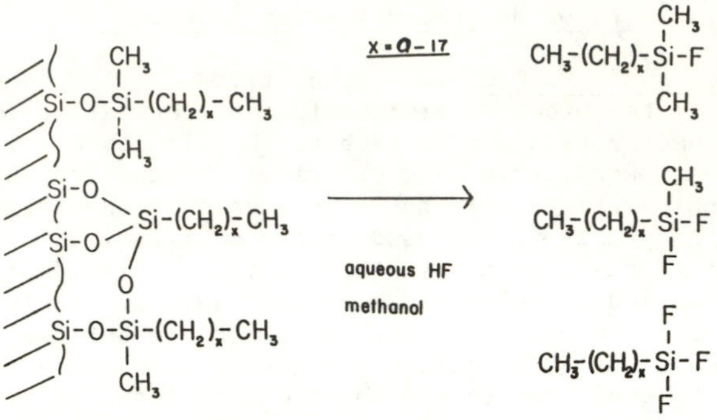

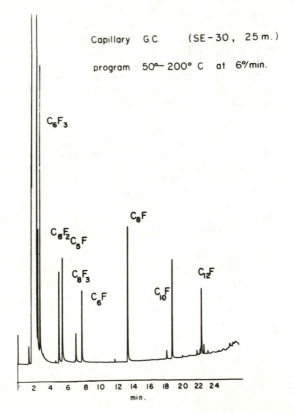

Figure 5. A simplified reaction scheme showing the major expected products of fluoride derivatization of bonded reverse phase silica HPLC packings.

Figure 6. Separation of a range of fluorosilyl derivatives on a capillary column, SE-30 stationary phase. Conditions: temperature program, 50-200°C at 6°C/mim. Peak identification is C_xF_3 represents a trifluoro-n-alkylsilane, C_xF_2 a difluoro-n-alkylmethylsilane, and C_xF a fluoro-n-alkyldimethylsilane.

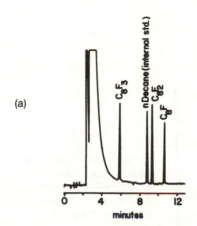

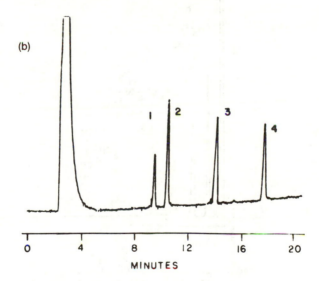

Figure 7. Optimized separation conditions for both octyl (top) and octadecyl (bottom) bonded phases on an SE-30 capillary (12 m) column. The octyl separation conditions are 65°C to 135°C at 6°C/min. The octadecyl separation conditions are 180°C to 210°C at 2°C/min. Identification of compounds in chromatograms are: (1) Fluorooctyldimethylsilane, (2) n-Decane (internal standard), (3) Difluorooctylmethylsilane, (4) Trifluorooctylsilane, (5) Octadecane (internal standard), (6) Trifluorooctadecylsilane, (7) Difluoromethyloctadecylsilane and (8) Fluorodimethyloctadecyl silane.

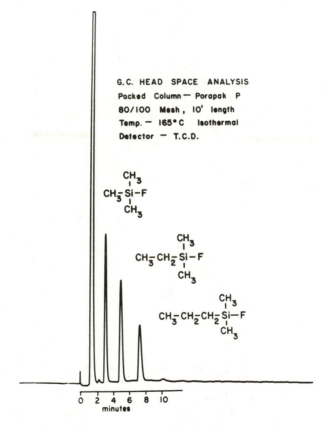

Figure 8.   GC head space analysis and separation of the methyl-,
ethyl- and n-propyldimethylfluorosilyl derivatives of several
silanes.   For analysis of "capping" agent trimethylsilane,
ethyldimethyl-bonded silica is added as the internal standard.
Chromatographic conditions: Column; Poropak P (3 m), Isothermal
at 165°C.

never exceeded 40-60% of the coverage calculated from elemental
analysis. Two major peaks were observed in the GC chromatograms
when analyzing trifunctional phases, as shown in Figure 10. The
relative areas of these peaks were consistent, and it was found
that 85% quantitative yield was observed when the areas of both
peaks were combined. Furthermore, under the GC conditions chosen,
no other peaks were observed, thus allowing for positive
identification of trifunctionally bonded phases. Quantitation was
possible using empirical calibration curves.

Confirmation by Mass Spectrometry. Extensive work was undertaken
to confirm the identity of all fluoroderivatives using mass
spectrometry. Electron impact fragmentation patterns confirmed
the identity of the mono-, di- and trifluoro derivatives of the
octyl and octadecyl ligands and all of the species labeled in
Figure 5.
     The problem of incomplete recovery of the trifluoro-
derivatives was found due to several competing side reactions of
these silanes. Four of the major species identified for the
trifunctional C_8 ligands are shown in Figure 11. Identities of the
two major species, accounting for 85% of the products formed, are
the trifluorosilane and the difluoromethoxy silane, shown in the
chromatogram of Figure 10. Extensive investigations of
acetonitrile as a wetting agent instead of methanol produced new
multiple products for the trifluorosilanes. It should be mentioned
that the three last species are thermodynamically less favored
than the first (12). Investigation of a wide range of reaction
temperatures and acid concentrations failed to eliminate completely
the formation of side-products. Thus, after extensive
investigations, the proposed methods were adopted for the
quantitative analyses for mono- and di- functional silanes of from
C_1 to C_{18} chain length, and qualitative and semi-quantitative
analysis of trifunctional bonded silanes of the same lengths.

Method Validation. The accuracy of the proposed gas
chromatographic method was validated against elemental analysis
(%C) over a wide range of bonding densities for monofunctional
methyl (end-capping), octyl and octadecyl silanes. All phases for
comparison were laboratory bonded using the appropriate
monochlorosilane ligand, and the octyl and octadecyl were not
end-capped. As shown in Table III, the regression slopes of the GC
and elemental methods were close to unity for both methyl and octyl
monofunctional ligands. Precision of the overall method for
monofunctional bonded silanes, including digestion and GC analysis,
was found to average 2-3% RSD for the methyl, 3-4% RSD for the
octyl phases, and slightly higher for the octadecyl phases, at
4-5%. It was found that injection port temperature was a critical
factor in the analysis of C_{18} phases. Temperatures below 240°C
resulted in injection port discrimination of the internal standard
and octadecyl fluoro derivatives.

Application to Commercial Phases and Silane Reagents. The proposed
GC method was applied to a number of commercial bonded phases, and

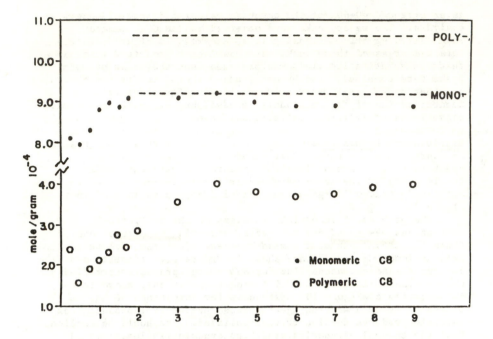

Figure 9.   Concentration of bonded silane determined as a
function of reaction time for the fluoride derivatization of both
monomeric and polymeric octyl bonded phases.   The dashed line
represents 100% recovery as determined by elemental analysis.

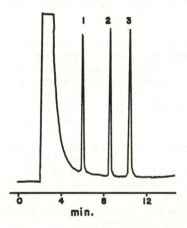

Figure 10.   Capillary GC chromatogram of a packing material
bonded with an octyltrichlorosilane.   Peak identities: 1.
Octyltrifluorosilane; 2.   n-Decane (internal standard); 3.
octyl-O-methoxydifluorosilane.

to commercial silane reagents used in bonding. Figure 12 shows an example of a commercial chlorodimethyldodecylsilane reagent. Significant contamination with C_8 and C_6 chain lengths is evident. This type of contamination is most likely due to impure olefin feedstocks from which the monochlorosilanes are synthesized via hydrosilation (18).

Tables IV and V summarizes the bonding densities of a variety of C_8 and C_{18} commercial phases along with the amount of trimethylfluorosilane (capping agent). The data shown were weight corrected back to grams of silane per gram of bare silica gel, according to Equation 1:

$$Y_i' = \frac{Y_i}{1 - \sum_{i=1}^{n} Y_i(MW)_i + \sum_{i=1}^{n} Y_i(MW)_H} \tag{1}$$

where Y_i' is the moles of ligand species i per gram of bare silica gel, corrected for the mass of all bonded species present, and Y_i is the original moles of ligand per gram of silica, uncorrected. MW_i is the molecular weight of the Y_{ith} bonded silane, not including the oxygen atom of the siloxane bond, and MW_H is the molecular weight of a hydrogen atom. Equation 1 is basically an extension of the equation of Unger et al (1), expanded to include now all chemical species present. As opposed to elemental analysis, the specificity of the GC method makes possible such analyses for multiple ligands, thereby providing a more complete picture of the bonded surface.

It is interesting to note that the majority of manufacturers are using monofunctional reagents for the octyl phases. The particular batches of LiChrosorb shown were the only phases found to use difunctional bonding. The GC method also correctly identified the trifunctional silane chemistries used in two commercial phases.

The ligand loadings shown in Table IV would seem at first to vary widely. However, if the loadings are divided through by the surface areas of the parent silica gel, the uniformity of the various commercial materials becomes more apparent. The data of Wise et al (58) were used for the BET areas, or when not available, nominal surface areas of the parent silica gels as issued by the manufacturers were used. It can be seen that the bonding densities, in terms of micromoles m^{-2}, are reasonably similar among most of the monofunctional octyl phases, and among the monofunctional octadecyl phases. The octadecyl phases were found bonded to a lower density than the octyl phases, an observation consistent with that found by Berendsen and others (57).

Conclusion. In conclusion, a simple, rapid set of GC methods based upon hydrofluoric acid digestion and either hexane extraction or head space analysis have been developed which will allow any laboratory equipped with a GC to identify and quantify the primary ligands and end-capping species present on the most widely used reversed-phase packing materials, i.e., octyl and octadecyl bonded phases. The method has been confirmed against elemental analysis,

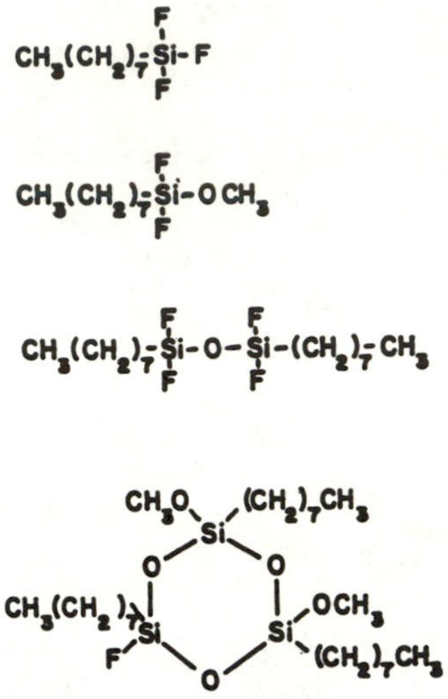

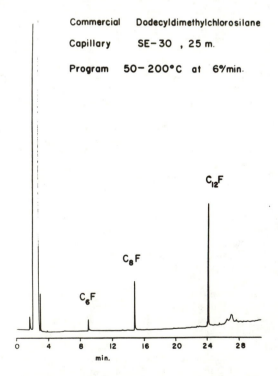

Figure 11. The structures of species identified by GC–MS in the HF digest of polymeric octyl bonded phases. The upper two products were the major components, comprising about 85% of the total silane.

Figure 12. A commercial dodecyldimethylchlorosilane after fluoride derivatization showing chain length contamination.

Table III.   Method Validation

Functionality	Phase	GC Method[a]	Elemental Analysis[a]
Monomeric	Octyl	10.76 (0.45)	10.87
		6.56 (0.04)	6.88
		5.53 (0.20)	5.91
		2.53 (0.21)	2.91
		%RSD (GC) = 2.1%	avg. diff. = 4.37% [b]
Monomeric	Octadecyl	6.60 (0.27)	6.62
		4.96 (0.12)	4.80
		3.31 (0.04)	3.20
		3.18 (0.08)	3.14
		1.54 (0.14)	2.45
		%RSD (GC) = 3.9%	avg. diff. = 2.97% [b]
Monomeric	Trimethyl	15.70 (0.11)	14.83
		5.56 (0.12)	5.75
		2.99 (0.03)	2.60
		2.36 (0.07)	2.34
		1.16 (0.02)	1.63
		%RSD (GC) = 1.7%	avg. diff. = 2.21% [b]

Phase	Y Intercept[c]	Correlation Coefficient	Slope
Octyl	−0.53 (0.01)	0.999	1.04 (0.08)
Octadecyl	−0.62 (0.14)	0.979	1.12 (0.58)
Trimethyl	−0.27 (0.04)	0.998	1.07 (0.26)

a. All data are in moles/gram x 10^4
    mean and (standard deviation)
b. Average difference as s_x/Mean x 100.
c. Elemental Data was given as the X axis and G.C. data were
    plotted on the Y axis.

Table IV. Commercial Octyl Phase

Company	Batch	Type[a]	BET[b]	Octyl[c]	Trimethyl[c]
Shandon MOS-Hypersil (5 um)	10/899	M	168	5.32 (0.33)	ND
Whatman Partisil-5	100645	M	347	4.85 (0.10)	11.40 (0.66)
Dupont (8 um)	23723-46	M	356	11.16 (0.51)	2.34 (0.01)
LiChrosorb (5 um)	W1535	D	280	2.68 (0.06)	ND
LiChrosorb (10 um)	VV1417	D	280	5.05 (0.15)	ND
Apex (5 um)	14769	M	172	5.49 (0.20)	---
Nucleosil (10 um)	9111	T		XXX	ND
Spherisorb (5 um)	B20018	T		XXX	---
Beckman Ultrasphere (5 um)	256-08	M		4.46 (0.15)	1.45 (0.11)

a. Type denotes functionality of organochlorosilane used in
    bonding reaction, M- monochloro, D- dichloro, T- trichloro.
b. BET numbers (m^2/g) from ref. 19, or from manufacturers data.
c. Octyl and Trimethyl concentrations are in moles/gram x 10^{-4}.
    Mean and (standard deviation) are listed, and are weight
    corrected to grams of ligand per gram of bare silica.

Table V. Commercial Octadecyl Phase

Company	Batch	Type	Octadecyl	Trimethyl
Dupont (12 um)		M	6.19 (0.17)	ND
LiChrosorb 10(um)	VV1219	D	5.50	ND
Nucleosil (10 um)	1041	T	XXX	3.24 (0.09)
Alltech (5 um)	19304	T	XXX	8.91 (0.19)
Beckman Ultrasphere IP		M	4.15 (0.11)	2.01 (0.18)
Adsorbo- sphere HS (7 um)	0207	M	8.51 (0.56)	2.05 (0.01)
Vydac 201-HS	103062	M	6.71 (0.10)	4.63 (0.12)
Versapack (10 um)	038403	M	3.84 (0.22)	11.82 (0.67)
Techsil (5 um)	T-06-05	T	XXX	2.38 (0.19)

and requires only 5-100 mg of packing material, depending on
bonding density. No special apparatus is necessary, and total
digestion and analysis time is about 3 hours at room temperature,
with multiple simultaneous sample preparation being possible.
While hydrofluoric acid must be used this reagent is no more toxic
than many other chemicals routinely handled in the laboratory.
Nevertheless, normal safety precautions for HF should be observed.

Acknowledgments   This work was supported by the National Science
Foundation, grant # CHE 8100224, and by the American Chemical
Society, grant # 16161-AC3, and by the Rutgers Research Council.

Literature Cited

1.   H. Engelhardt and H. Muller, J. Chromatogr. 218, 395 (1981).
2.   N. Tanaka, Y. Tokuda, K. Iwaguchi and M. Araki, J. Chromatogr.
     239, 761 (1982).
3.   S.A. Wise and W.E. May, Anal. Chem. 55, 1479 (1983).
4.   R.J. Amos, J. Chromatogr. 204, 469 (1981).
5.   A. Sokolowski and K.G. Wahlund, J. Chromatogr. 189, 299
     (1980).
6.   M. Verzele, M. De Potter and J. Ghysels, J. High Resolut.
     Chromatogr. Commun. 2, 151 (1979).
7.   M. Verzele, J. Lammens and M. van Roelenbosch, J. Chromatogr.
     186, 435 (1979).
8.   G.E. Berendsen, K.A. Pikaart and L. De Galan, J. Liq.
     Chromatogr. 3, 1437 (1980).
9.   B.A. Bidlingmeyer, J.K. Del Rios and J. Korpl, Anal. Chem. 54,
     442 (1982).
10. N.H. Cook and K. Olsen, J. Chromatogr. Sci. 18, 512 (1981).
11. L. Nondek and V. Vyskocil, J. Chromatogr. 206, 581 (1981).
12. W.R. Melander, J. Stoveken and C. Horvath, J. Chromatogr. 199,
     35 (1980).
13. P. Roumeliotis and K.K. Unger, J. Chromatogr. 149, 211 (1978).
14. K. Karch, I. Sebastian and I. Halasz, J. Chromatogr. 122, 3
     (1976).
15. E.J. Kikta and E. Grushka, Anal. Chem. 48, 1098 (1976).
16. H. Hemetsberger, P. Behrensmeyer, P. Henning and H. Ricken,
     Chromatographia 12, 71 (1979).
17. C.T. Wehr, L. Correis and S.R. Abbott, J. Chromatogr. Sci. 20,
     114 (1982).
18. C.H. Lochmuller, A.S. Colborn and M.L. Hunnicutt, Anal. Chem.
     55, 1344 (1983)
19. R.L. Gilpin and J.A. Squires, J. Chromatogr. Sci. 19, 195
     (1981).
20. R.K. Gilpin, M.E. Gangoda and A.E. Krishen, J. Chromatogr.
     Sci. 20, 245 (1982).
21. G.E. Berendsen and L. Galan, J. Liq. Chromatogr. 1, 561
     (1978).
22. H. Hemetsberger, P. Behrensmeyer, J. Hening and H. Ricken,
     Chromatographia 12, 71 (1979).
23. I.S. Krull, M.H. Wolf and R.B. Ashwort, Int. Lab. July/Aug.,
     25 (1978).
24. J.B. Crowther and R.A. Hartwick, Chromatographia 16, 349
     (1982).

25. J.B. Crowther, S.D. Fazio and R.A. Hartwick, J. Chromatogr.
    (Baden Baden Proc., in press (1983)).
26. R. Roumeliotis and K.K. Unger, J. Chromatogr. 142, 213 (1978).
27. P.A. Bristow and J.H. Knox, Chromatographia 10, 279 (1977).
28. C.J. Little, A.D. Dale and M.B. Evans, J. Chromatogr. 153, 381
    (1978).
29. K.K. Unger, N. Becker and P. Roumeliotis, J. Chromatogr. 125,
    115 (1976).
30. L.C. Sander, J.B. Callis and L.R. Field, Anal. Chem. 55, 1068
    (1983).
31. R.P.W. Scott, and S.J. Traiman, J. Chromatogr. 196, 193
    (1980).
32. H. Hemetsberber, W. Massfeld and H. Ricken, Chromatographia 9,
    303 (1976).
33. J.L.M. van de Venne, J.P.M. Rindt, G.J.M.M. Coenen and
    C.A.M.G. Cramers, Chromatographia 13, 11 (1980).
34. R.E. Majors and M.J. Hopper, J. Chromatogr. Sci. 12, 767
    (1974).
35. L.T. Zhuravlev, A.V. Kiselev and V.P. Naidina, Russ. J. Phys.
    Chem. 42, 1200 (1968).
36. C.H. Lochmuller, S.F. Marshall and D.R. Wilder, Anal. Chem.
    52, 19 (1980).
37. G.E. Mariel, D.W. Sindorf and V.J. Bartuska, J. Chromatogr.
    205, 438 (1981).
38. R.K. Gilpin and M.E. Gangoda, J. Chromatogr. Sci. 21, 352
    (1983).
39. S.G. Weber and W.G. Tramposch, Anal. Chem. 55, 1771 (1983).
40. M. Verzele, P. Mussche and P. Sandre, J. Chromatogr. 190, 331
    (1980).
41. L. Hansson and L. Trojer, J. Chromatogr. 207, 1 (1981).
42. R.H. Prince in M.T.P. International Review of Science:
    Inorganic Chemistry Series I, Vol. 9 (ed. M.L. Tobe), 1972,
    p. 353.
43. M.L. Miller, R.W. Linton, S.G. Bush, J.W. Jorgenson, Anal.
    Chem. 56, 2204-2210 (1984).
44. J.B. Crowther, S.D. Fazio, R. Schiksnis, S. Marcus and R.A.
    Hartwick, J. Chromatogr. 289, 367-375 (1984).
45. J.B. Crowther and R.A. Hartwick, Chromatographia 16, 349
    (1982).
46. J.B. Crowther, S.D. Fazio and R.A. Hartwick, J. Chromatogr.
    282, 619 (1983).
47. E.J. Corey and A. Venkateswarlu, J. Am. Chem. Soc. 94, 6190
    (1972).
48. E.J. Corey and B.B. Snider, J. Am. Chem. Soc. 94:7, 2549
    (1972).
49. P.J. Holmes, "The Electrochemistry of Semiconductors",
    Academic Press (1962).
50. J.P. Erard and E. Kovats, Anal. Chem. 54, 193-202 (1982).
51. B.J. Aylett, "Organometallic Compounds", Vol. 1, Part 2, John
    Wiley & Sons, New York (1979).
52. L.H. Sommer, "Stereochemistry, Mechanism and Silicon", McGraw
    Hill (1965).
53. J.A. Gardella, J.S. Chen, J.H. Magill, and D.M. Hercules,
    J. Am. Chem. Soc. 105, 4536-4541 (1983).

54. N. Okui, J.H. Magill and K.H. Gardner, J. Appl. Phys. 48, 4116
     (1977).
55. N. Okui, H.M. Li and J.H. Magill, Polymer 18, 1152 (1977).
56. N. Okui and J.H. Magill, Polymer 19, 411 (1978).
57. G.E. Berendsen, K.A. Pkaart, L. De Galan, J. Liquid
     Chromatogr. 3, 1437 (1980).
58. L.C. Sander and S.A. Wise, personal communication.

RECEIVED October 3, 1985

3

Column-Packing Structure and Performance

Vladimir Rehák

Laboratory of Infectious Diseases, Faculty of Pediatrics, Charles University, Prague, Czechoslovakia

The significance of bonded stationary phases is documented by constantly growing number of publications dealing with them, and from the use of a new expression "BPLC" [Bonded Phase Liquid Chromatography].

During the period of development of these useful packing materials many different preparation procedures were used and thoroughly studied. The method using the reaction between silica and haloalkylsilane [alkoxyalkylsilane alternatively] has been found the best one and it is, nowadays, used exclusively. Although these packings are widely used, their structure and mainly the separation mechanism on them are still not completely understood. The general conclusion has been that the separation mechanism on these packings is very complicated involving a number of processes and has been valid up to now. The situation is simpler on non-polar phases [reversed-phase packings], where the bonded alkyls contain only $-CH_2-$ and $-CH_3$ groups. Such materials are, therefore, studied more frequently.

Our attention has been focused on these packings too, and only one factor influencing retention on non-polar chemically bonded stationary phases in HPLC--namely the bonded phase structure--was studied. Two groups of packings [monomeric, "bristle" type] were synthesized [detailed description in 1,2], one having branched structures, the other unbranched ones. Materials having either the same total number of C atoms in the bonded alkyl or the same number of C atoms in the longest chain of bonded alkyl were compared. Packings giving reproducible results are listed in TABLE I.

Experimental

Preparation of packings. Commercially unavailable silanes were synthesized by the reaction described by Barry and coworkers [3] [1-olefin reacts with silane under this scheme]:

$$R-CH=CH_2 + H-Si-R_xCl_y \longrightarrow R-CH_2-CH_2-Si-R_xCl_y,$$

where R is an alkyl of required structure, Rx is methyl group, and x+y=3. Resulting haloalkylsilanes were purified by vacuum

0097–6156/86/0297–0056$06.00/0

distillation and characterized by determination of chlorine content, refractive index value and C,H content.

Available silanes were purified and characterized prior to reaction with silica-gel in the same manner.

Silica-gel [LiChrosorb SI 100 - Merck, particle size 10 μm irregularly shaped] was acid washed with the mixture of sulfuric and nitric acids 9:1 at the temperature of 130 degrees of centigrade. When the washing with water reached neutrality, it was dried in vacuo overnight at 150 degrees of centigrade. Its specific surface area and pore size distribution were then established. This material was then reacted with appropriate silane in toluene [modified method of Kirkland and deStefano 4]. The reaction mixture was heated under reflux cooler and the equimolar amount [to the amount of silane] of triethylamine was added [Schmidt 5]. The reaction was completed within three hours, as determined by C,H analysis of samples taken from the reaction mixture every half an hour. No difference between branched and unbranched packings were observed.

Characterization of packings. I. C,H analysis: Thoroughly washed and dried samples were taken for the estimation of carbon content. All materials were then end-capped with hexamethyldisilazane in toluene, washed and dried again and new samples of packings were taken for the C,H analysis. Carbon content of all packings increased slightly by the end-capping [2-7 percent of the initial value]; once more with no significant difference between branched and unbranched packings. TABLE II summarizes the results.

II. Specific surface area measurements: The method of thermal desorption introduced by Nelsen and Eggertsen [6] was employed using apparatus by Grubner [7]. The results in TABLE III show that the specific surface area decreases by bonding of an alkyl. This decrease ranges between 35 and 52 percent of the value for unmodified silica for the packings described in this study. Values above 50 percent were obtained for bonded n-octadecyl and for silica modified with HMDS directly. For all other packings, values between 35 and 52 percent were found. The difference in decrease of specific surface area between branched and corresponding unbranched phase is small. In all cases the values are slightly lower for branched packings; however, the difference is comparable within the accuracy of the method.

III. Pore size distribution: Our assumption that bonding of an alkyl reduces the main pore diameter was confirmed by Martin's method [8]. The decrease in this value is in a very good agreement with the value calculated for each packing, using the simplest possible model--a cylinder, to whose inner walls maximum stretched alkyls are bonded. TABLE IV lists the calculated and measured values. The results for 1-ethyladamantyl cannot be correlated with the theory, apparently because of its structure differs greatly from those of the other alkyls. It cannot be excluded that only this structure makes possible specific interactions leading to the deviation observed. [Martin's method is based on gel-permeation principle and the specific interactions must be excluded. It uses, as the test solutes, polystyrenes of known size and it is possible that polystyrene molecule not only enters the pores, as required, in part specifically interacts with the structure of bonded adamantyl,

TABLE I – List of prepared packings.

Bonded alkyl	Abbreviation	C atoms in chain	total	Reference packing
2,4,4-trimethylpentyl	2,4,4	5	8	n-pentyl, n-octyl
4-butyloctyl	C_8--	8	12	n-octyl, n-dodecyl
1-ethyladamantyl	adam		12	n-dodecyl
n-butyl	C_4	4	4	
n-hexyl	C_6	6	6	
n-octyl	C_8	8	8	
n-dodecyl	C_{12}	12	12	
n-dodecyl	C_{18}	18	18	
methyl	HMDS	1	1	

TABLE II – C,H analysis results.

Packing	Carbon content before end-capping	Carbon content after end-capping	Increase
2,4,4	9.29% by weight	9.53%	2.6% rel.
C_8--	11.74	12.40	5.6
adam	12.48	12.88	3.1
C_8	10.88	11.13	2.3
C_{12}	12.43	13.34	6.8
C_{18}	16.32	16.79	2.8

TABLE III - Specific surface area measurements.

Packing	%C	Amount of silica in 1 g of packing	Corresponding surface area	Area after bonding	Decrease
2,4,4	9.29	0.8633 g	292.0 m^2	166.5 m^2	42.98%
C_8	11.74	0.8391	281.3	180.3	35.93
adam	12.48	0.8341	279.7	281.3+	
HMDS	3.74	0.9526	314.0	149.8	52.29
C_8	10.88	0.8400	286.0	170.3	40.45
C_{12}	12.43	0.8300	277.7	162.0	41.60
C_{18}	16.32	0.7874	264.0	175.1	52.65

Note: Carbon contents values before end-capping are given.

TABLE IV - Pore size distribution - the mean pore diameter.

Packing	Approx. length of alkyl chain Å	Mean pore diameter Å	
		Calculated	Measured
2,4,4	3.6	90.0	91.2
C_8--	8.4	83.0	84.1
adam	8.0	84.0	120.0+
C_8	8.4	83.0	86.6
C_{12}	13.2	74.0	78.0
C_{18}	20.0	60.0	65.7

thus increasing elution volume of entire polystyrene; in this case, however, false results will be calculated.]

IV.   Surface concentration of bonded molecules.   Based on data given in TABLE III, the values of surface concentrations of bonded molecules were calculated.   The results presented in TABLE V show that all packings in the present work were prepared   under the condition of maximum coverage [9].   The site requirements of bonded molecules were found to be between 40 and 52 square angstromes, roughly corresponding to the data reported by Hemetsberger [10]. The reliability of these data is strongly dependent on the accuracy of both surface area measurements and C,H analyses.   A comparison of values for pairs of the corresponding branched and unbranched packings indicates that the coverage of branched phases is slightly lower.

Chromatography. I.   Packing of columns: Balanced density slurry method was employed and the quality of packed columns was tested chromatographically in methylene chloride.   Benzene was injected as testing solute.   Columns which gave satisfactory results [i.e., symmetrical elution curve, relative band broadening of 50 µm at linear velocity   of 5 mm/sec, and permeability better than $1.10\pm$ -9 cm±2] were used for further experiments.

Loadability of columns estimated as described by Karch [11], was found to be comparable for both branched and unbranched packings. Values between 5.10 ± -4 and 1.10 ± -5 g per gram of packing were found, depending on carbon content of the packing and solute used for this estimation.

The rate of equilibration after a change in the mobile phase composition is also comparable for both types of packings.   If the methanol content in a new mobile phase is greater than 1 percent, constant capacity factors are reached very quickly--after 15–20 ml have been pumped through the column.   If an equilibrium cannot be reached even after several litres of pure water have been pumped through the column suggests non-polar phase not wettable by water.

The excellent reproducibility of packing and the efficiency of prepared columns were the same for both branched and unbranched packings and no differences were found in their loadability.   These facts enable us to compare different bonded phases and to interpret the influence of bonded phase structure on retention.

II.   Mobile phases: Three water/methanol mixtures of different composition [98:2, 70:30, 30:70] and pure methanol were used as mobile phases possessing different elution strengths.

III.   Solutes.   At least three groups of solutes were used in each of four mobile phases described above.   It was desirable for the solutes within these groups to have k' values between 0.1 and 15,   to create homologous series or to have very similar structure. If possible, each group should be eluted within reasonable time at least in two mobile phases used in this study.

IV.   Retention.   The dependence of retention described in terms of k' on several parameters [the length and structure of bonded molecule, size and structure of solute molecule, and the mobile phase elution strength] was studied.   The measured k' values were normalized to 1 µmol of a bonded molecule in order to enable the structure effect on retention to be determined with greater sensitivity.   Figures 1 and 2 show examples of dependence of log k'

TABLE V - Surface concentration of bonded molecules.

Packing	μmol of silane bonded to 1 m^2	Number of silane molecules per 100 Å	Site requirement $Å^2$
2,4,4	3.40	2.05	48.87
C_8--	2.94	1.77	56.38
adam	3.15	1.90	52.66
C_8	4.09	2.46	40.61
C_{12}	3.15	1.90	52.66
C_{18}	2.90	1.75	57.11

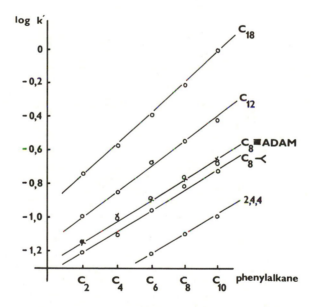

Figure 1.   Influence of the bristle length on the retentions of phenylalkanes. Eluent: methanol. Samples: ethylbenzene, butylbenzene, hexylbenzene, phenyloctane and phenyldecane. Stationary phases as given in TABLE I.

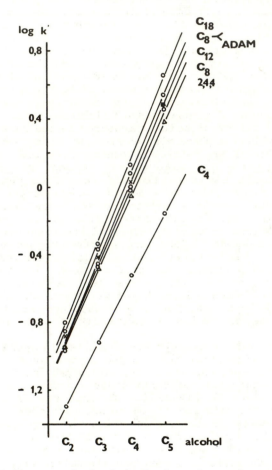

Figure 2.   Influence of the bristle length on the retentions of
n-alcohols. Eluent: water/methanol 70/30. Samples:
ethanol, propanol, butanol and pentanol. Stationary
phases as given in TABLE I.

on the length of bonded molecule [on bristle length]. For branched packings, the length of bonded molecule is represented by that of the longest chain. For both types of phases, the k' value is proportional to the bristle length.

The following differences in retention were found. For a given solute, the k' value and especially the ratio of k' on branched to k' on the corresponding unbranched packings are determined by mobile phase elution strength and solute molecule structure. In mobile phases possessing high elution strength the k' values for given solute are always higher on unbranched packings. This difference decreases with decreasing elution strength and the ratio of k' values can be reversed for many solutes and the retention on the branched phases is greater than on the corresponding unbranched phases. The solute structure seems to be the principal factor determining the conditions under which this reversal of the retention occurs. Figures 3 and 4 illustrate this. They show the retention of different solutes measured in the same mobile phase [influence of mobile phase elution strength is excluded]. While, in all cases, methyl derivatives of benzene have lower k' on branched than on corresponding unbranched phases, the retention of dialkylphthalates is quite different from phase to phase. Dialkylphthalates retention on bonded 2,4,4-trimethyl pentyl and 4-butyl octyl is comparable with that on the corresponding unbranched packings. The retention of these solutes on the bonded 1-ethyl adamantyl is greater than on n-dodecyl [corresponding unbranched phase] or on n-octadecyl [the largest unbranched phase].

The special selectivity of the bonded phases for a solute the same structure as these, was also studied. The bonded 1-ethyl adamantyl was chosen as striking example. The relative retention of adamantane to n-decane in methanol was measured. This value was found to be lower than 1 on all packings except the bonded adamantyl. On this packing the retention of both C10 hydrocarbons is equal.

The last part of experiments consisted in application of the prepared packings [branched] to the separation of several drug mixtures. Barbiturates, anabolic steroids and anticonvulsants were chosen. Chromatograms of barbiturates obtained on the branched phases are shown in Figure 5. In this set of ten different barbiturates the main problem is to separate phenobarbitone from allobarbitone [on the unbranched packings]. Very good selectivity of all branched packings for these two barbiturates is evident but overlapping peaks for other compounds appeared. As to the barbiturates the branched phases have the selectivity different from unbranched ones. Peak shape can be improved significantly if using alkalized mobile phase, but this shortens the column life drastically and so this modification was not employed with the unique branched packings.

Separation of anabolics is not very difficult. Good separation of testosteron and its methyl derivative [especially for the doping tests] can be achieved on the unbranched packings. The use of branched packings didn't bring any improvement in this respect and separation of other compounds was even worse.

The shifts in retention of anticonvulsants on all tested packings were not significant and both types of packings were successfully used for the analysis of serum samples from epileptic patients.

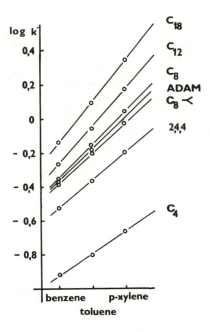

Figure 3.   Influence of the solute molecule size and structure on
            the retentions of benzene methyl derivatives. Eluent:
            water/methanol 30/70.   Samples: benzene, toluene and
            p-xylene.   Stationary phases as given in TABLE I.

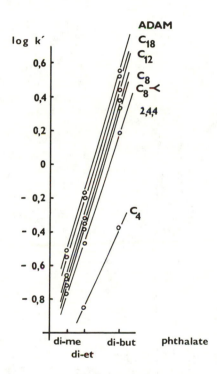

Figure 4.   Influence of the solute molecule size and structure on
the retentions of dialkyl phthalates.   Eluent:
water/methanol 30/70.   Samples: dimethyl, diethyl and
dibutyl phthalate.   Stationary phases as given in
TABLE I.

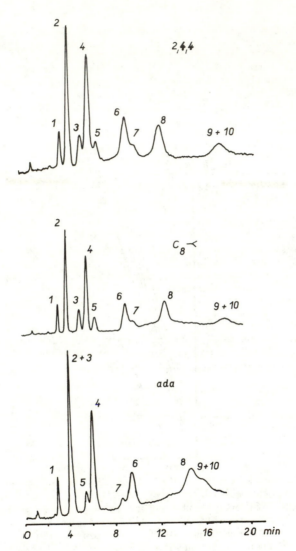

Figure 5. Separation of barbiturates on the branched phases.
Eluent: water/methanol 65/35. Samples: 1-barbitone
(5,5-diethyl barbituric acid), 2-heptobarbitone (5-
methyl-5-phenyl), 3-allobarbitone (5,5-diallyl), 4-
phenobarbitone (5-ethyl-5-phenyl), 5-aprobarbitone (5-
allyl-5-isopropyl), 6-cyclobarbitone (5-ethyl-5-cyclo-
hexenyl), 7-butobarbitone (5-ethyl-5-butyl), 8-
hexobarbitone (5-methyl-5-cyclohexenyl N-methyl
barbituric acid), 9-amobarbitone (5-ethyl-5-isopentyl
barbituric acid), 10-pentobarbitone (5-methyl-5-
pentyl). Stationary phases as given in TABLE I.

Conclusion

The difference observed in retention on the branched and unbranched phases are caused exclusively by the structure of bonded molecule because other factors can be neglected. The basic features of packings of both types, i.e. carbon content, pore size distribution, coverage, specific surface area, the quality of columns packed with these phases, loadability and rate of equilibration are similar.

The difference in retention on phases of different structures are more pronounced in mobile phases of lower elution strength. The k' values are governed by the solute structure. The results suggest that comparison of retention on the branched and unbranched phases could also be used for solving more general problems of the separation mechanism, based on study of molecular interactions. The phases with branched structure could also be useful in extending the application of HPLC.

Acknowledgments

The author is grateful to S.E.A. for the support of this work, to all members of Institut für Angewandte Physikaliche Chemie, Universität des Saarlandes, Saarbrücken, W. Germany (prof. I. Halasz, Head), where the main part of experiments were carried out, to all members of Laboratorium voor Instrumentele Analyse, Technische Hogeschool Eindhoven, Eindhoven, The Netherlands, (Dr. Ir. C. A. Cramers, Head) for the effective cooperation, and to Drs. V. Sváta and B. Máca, Department for Organic Chemistry, Faculty of Science, Charles University, Prague, Czechoslovakia, for their help in the synthesis of unavailable olefins.

Literature Cited

1. V. Rehák, PhD Thesis, Charles University, Prague, Czechoslo-vakia 1980
2. V. Rehák, E. Smolková, J. Chromatogr., 191 (1980) 71
3. A. J. Barry, L. dePree, J. W. Gilkey, D. E. Hook, J. Amer. Chem. Soc., 69 (1947) 2916
4. J. J. Kirkland, J. J. deStefano in Advances in Chromatography 1970, A. Zlatkis editor, Chromat. Symp. Dept. of Chem., University of Houston, Texas, USA
5. H. Schmidt, Doktorarbeit, Universität des Saarlandes, Saarbrücken, W. Germany 1979
6. F. M. Nelsen, F. T. Eggertsen, Anal. Chem., 30 (1958) 1387
7. O. Grubner, Z. Phys. Chem., 216 (1961) 287
8. K. Martin, Doktorarbeit, Universität des Saarlandes, Saarbrücken, W. Germany 1975
9. K. Unger, Angew. Chem., 84 (1972) 331
10. H. Hemetsberger, W. Maasfeld, H. Ricken, Chromatographia 9 (1976) 303
11. K. Karch, Doktorarbeit, Universität des Saarlandes, Saarbrücken, W. Germany 1976

RECEIVED October 16, 1985

4

Differences in Selectivity of Reversed-Phase Columns for High-Performance Liquid Chromatography

I. Wouters, S. Hendrickx, E. Roets, J. Hoogmartens, and H. Vanderhaeghe

Katholieke Universiteit Leuven, Instituut voor Farmaceutische Wetenschappen, Laboratorium voor Farmaceutische Chemie, Van Evenstraat 4, B-3000 Leuven, Belgium

Results obtained by high performance liquid chromatography of fifteen cephalosporins on eight brands of C8
and C18 packing materials are reported. A number of
characters of the packing materials and of the columns
prepared with them are compared, i.e. surface, pore diameter, carbon content, methyl red adsorption values,
plate number and separation factor. Important differences in selectivity are observed between manufacturers,
but also between batches from the same manufacturers.
The correlation between the selectivity and the column
characters considered is poor.

In most papers the selectivity of the column packing used, compared
with that of other packing materials of the same type, is not discussed.

Usually, when authors describe the separation of a given number
of products, a particular brand of reversed phase material is used
with a particular mobile phase. This makes it difficult to decide
whether differences in selectivity are due to changes in mobile phases or in packing materials. Therefore, one is often tempted to believe that differences in selectivity are mainly caused by the mobile phase. This is a misconception already mentioned by several
authors (1-9).

In this study the results, obtained by chromatography of fifteen cephalosporin antibiotics on different brands of C8 and C18
reversed phase materials, are compared. Important differences in
selectivity between manufacturers and also between batches from the
same manufacturer, are observed. Part of this study has already been
published (10).

Samples

The following cephalosporins were chromatographed : cephalosporin C
dihydrate (1), cefadroxil monohydrate (2), cefatrizine propylene-
glycolate (3), cefaloglycin dihydrate (4), cefaclor monohydrate (5),
cefalexin monohydrate (6), cefradine anhydrate (7), cefamandole (8),
cefalotin (9), cefoxitin (10), cefaloridine δ-form (11), cefapirin

0097-6156/86/0297-0068$06.00/0

(12), cefazolin (13), cefuroxime (14), cefotaxime (15). The structures are shown in Figure 1.

Stationary Phases and Columns

Information on the packing materials and the columns is summarized
in Table I. Except for the prepacked columns, all columns were
packed following the same procedure, which was not necessarily the
optimum way for each particular packing material. The columns were
packed using a slurry of 2.7 g packing material (3.2 g for Zorbax)
in 15 ml toluol-cyclohexanol 1:2, a Haskel pump Model DSTV-122 with
an inlet pressure of 5 bar and with methanol as the pressurizing
liquid.The column dimensions were 25 cm and 4.6 mm I.D., except for
the Hibar column: 25 cm x 4 mm I.D.
    All columns were checked by chromatography of a mixture of ben-
zene, naphthalene, phenanthrene and anthracene with methanol-water
(70:30) as the mobile phase. For Polygosil, methanol-water (60:40)
had to be used to obtain separation.
    The plate numbers per meter listed in Table I are not an abso-
lute indication for the intrinsic efficiency of the packing materials. Indeed, previous experience with other packing methods resulted in higher efficiency especially with LiChrosorb and µ Bondapak.
This explains the discrepancy between the LiChrosorb RP-8 10 µm
Hibar column and the corresponding home-made column. Although the
plate number affects resolution, it is not affecting the selectivity, which is our main interest. The separation factor, α , allows a
rough distinction between C8 and C18 derivatised silicas. The C18
materials do not necessarily show higher capacity factors (k') for
anthracene.
    Plate numbers, calculated on a well separated cephalosporin are
also summarized in Table I. This plate number can agree or widely
differ from the corresponding plate number obtained for naphthalene.
For Partisil and Zorbax extreme differences are observed. Generally
lower N/m are recorded for cephalosporins.

Chromatographic Conditions

The mobile phase was acetonitrile-water-0.2 M phosphate buffer pH
7.0. The percentage of organic modifier was varied in order to obtain complete elution within about 40 min, resulting in organic
modifier contents from 1 to 11.5 %. The flow rate was 1 ml/min and
the detector was set at 254 nm with a sensitivity of 0.05 A.U.F.S.
All separations were carried out at room temperature.

Results and Discussion

Figures 2 and 3 show a comparison between a chromatogram of LiChrosorb RP-8 10 µm obtained recently and another obtained a few years
ago. Differences in selectivity are apparent; the old packing material gives the best results, probably due to a somewhat lower
coverage and better interaction between polar solutes and free silanols. The older chromatogram was obtained at a somewhat lower flow
rate (0.8 ml/min) and a different chart speed. Previously the elution order of the cephalosporins at the end of the chromatogram was
13, 12, 4, and now it is 4, 13, 12. Before 5 and 7 coeluted and now

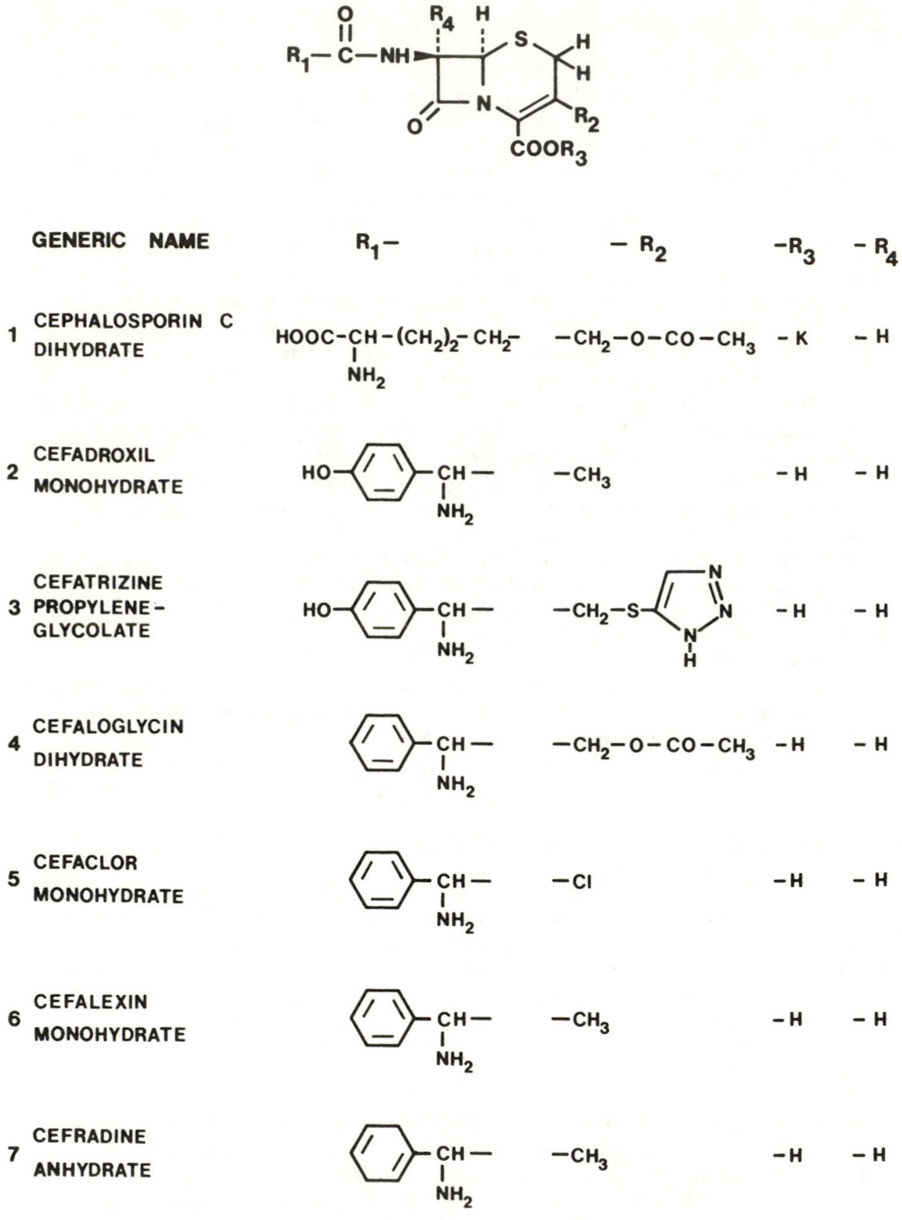

GENERIC NAME	R_1-	$-R_2$	$-R_3$	$-R_4$
1 CEPHALOSPORIN C DIHYDRATE	$HOOC-\underset{\underset{NH_2}{\vert}}{CH}-(CH_2)_2-CH_2-$	$-CH_2-O-CO-CH_3$	$-K$	$-H$
2 CEFADROXIL MONOHYDRATE	$HO-\bigcirc-\underset{\underset{NH_2}{\vert}}{CH}-$	$-CH_3$	$-H$	$-H$
3 CEFATRIZINE PROPYLENE-GLYCOLATE	$HO-\bigcirc-\underset{\underset{NH_2}{\vert}}{CH}-$	$-CH_2-S-\underset{H}{\overset{N=N}{\underset{N}{\bigcirc}}}$	$-H$	$-H$
4 CEFALOGLYCIN DIHYDRATE	$\bigcirc-\underset{\underset{NH_2}{\vert}}{CH}-$	$-CH_2-O-CO-CH_3$	$-H$	$-H$
5 CEFACLOR MONOHYDRATE	$\bigcirc-\underset{\underset{NH_2}{\vert}}{CH}-$	$-Cl$	$-H$	$-H$
6 CEFALEXIN MONOHYDRATE	$\bigcirc-\underset{\underset{NH_2}{\vert}}{CH}-$	$-CH_3$	$-H$	$-H$
7 CEFRADINE ANHYDRATE	$\bigcirc-\underset{\underset{NH_2}{\vert}}{CH}-$	$-CH_3$	$-H$	$-H$

Figure 1.   Cephalosporin structures.   Continued on next page.

Figure 1 continued.   Cephalosporin structures

Table I : CHARACTERISTICS OF PACKING MATERIALS AND COLUMNS

Packing Materials	Shape Irregular (I) Spherical (S)	Plates per m determined on naphthalene (N/m)	Separation factor phenanthrene-anthracene	Capacity factor k', anthracene	Plates per m (N/m), determined on a well separated cephalosporin
LiChrosorb RP-8 10 μm Hibar (b)	I	19,600	1.09	6.2	14,000
LiChrosorb RP-8 10 μm	I	10,300	1.08	5.7	11,600
LiChrosorb RP-8 5 μm	I	26,800	1.09	7.1	25,200
Zorbax C8 7 μm	S	33,600	1.10	9.9	4,000
Polygosil C8 10 μm	I	11,600[a]	1.08[a]	6.8[a]	8,800
Nucleosil C8 10 μm	S	15,800	1.05	4.4	12,000
RSil C18 LL 10 μm	I	18,400	1.12	7.4	12,800
μBondapak C18 10 μm	I	7,200	1.12	9.2	4,000
Nucleosil C18 10 μm (b)	S	12,000	1.19	6.0	6,800
Partisil ODS 10 μm (b)	I	13,600	1.17	2.1	3,600

a : a different mobile phase was used   b : prepacked column

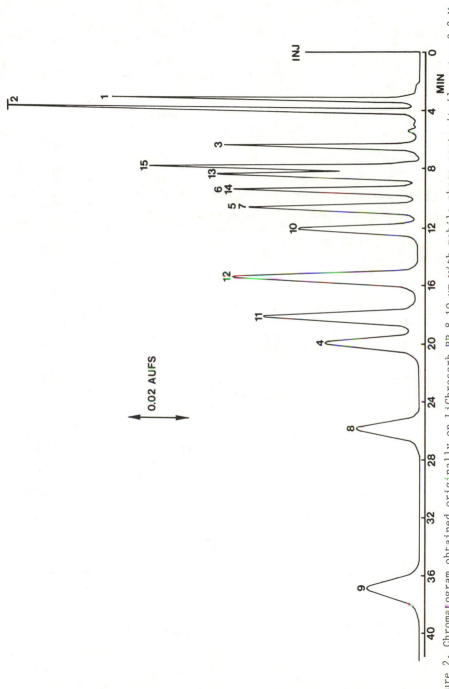

Figure 2. Chromatogram obtained originally on LiChrosorb RP-8 10 um with mobile phase acetonitrile–water–0.2 M phosphate buffer pH 7.0 (12:83:5).

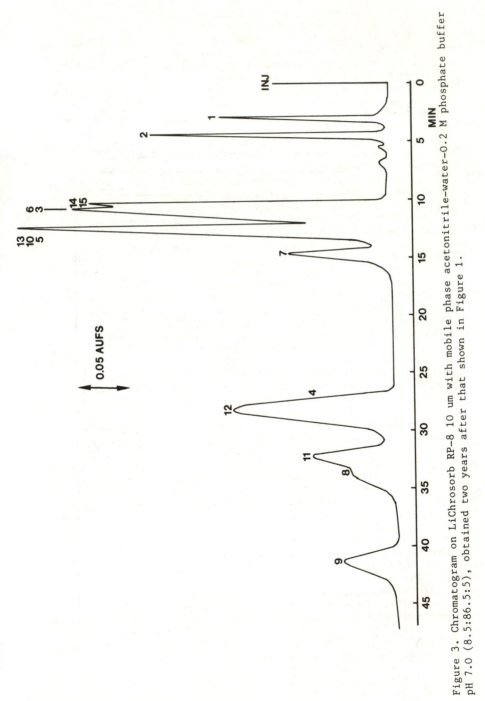

Figure 3. Chromatogram on LiChrosorb RP-8 10 um with mobile phase acetonitrile-water-0.2 M phosphate buffer pH 7.0 (8.5:86.5:5), obtained two years after that shown in Figure 1.

they are separated, 3 was baseline separated before 15 and 16 and it now elutes after 15 and 16, and 11 was well separated and it now elutes within a group. LiChrosorb materials not only showed important differences in selectivity, but also different packing procedures had to be used. However, the materials were always available under the same label. These results show that between different batches, selectivity differences exist and that a given separation is not necessarily reproducible from batch to batch of different ages.

The same phenomenon was observed when a commercially available LiChrosorb RP-8 10 µm Hibar (Fig. 4) and a corresponding home-packed column (Fig. 3) were compared. Although the same selectivity was expected, important differences were observed. Replacing a home-packed LiChrosorb RP-8 10 µm (Fig. 3) by a corresponding 5 µm column (Fig. 5), the same selectivity was expected since only particle size influences the efficiency. In fact a much different selectivity was observed and for the 5 µm column a higher percentage of acetonitrile was needed to elute the cephalosporins within the desired time (40 minutes). These results show that when smaller particles are introduced in order to decrease analysis time, important changes in selectivity can occur.

In Table II, the order of elution obtained on all the columns examined are listed. On the Zorbax column the cephalosporins are poorly resolved. However, from the N/m value for naphthalene (Table I) one would expect this column to be the best. This is probably due to the fact that it is a well covered material where less influence of residual silanols occurs. This poor resolution does not mean that Zorbax is an inferior packing material since we obtained very nice separations for other drugs, but it shows that a material which is excellent for many separation problems and has the highest claimed N/m value is not necessarily the best for a particular separation problem.

Using Nucleosil and Polygosil C8 from the same manufacturer, large selectivity differences are observed which can partly be due to the fact that the parent silicas are different. Generally, spherical particles give higher plate numbers than irregular ones. In spite of the higher plate number for Nucleosil, expected to give the better resolution, the less expensive Polygosil shows better selectivity. For Nucleosil C8, the same phenomenon as for LiChrosorb C8 has been observed : a chromatogram obtained years ago with another batch showed a different selectivity. Results obtained on Nucleosil C18 10 µm and µBondapak C18 10 µm again emphasize selectivity differences between manufacturers.

On the RSil and Partisil columns, cephalosporin 12 is strongly retained. This can be explained by the low loading of these columns which allows a very good interaction of this positively charged cephalosporin with the free silanols.

Generally, the position in the elution order can be quite variable for some cephalosporins : it can be seen that 10 moves from the third place to the tenth, while 7 is always found at the ninth or tenth place. It is impossible to see any relationship between the structure of the cephalosporins and their elution order which is different on all the columns. It is clear that it is impossible to make valid predictions about the selectivity of a column towards cephalosporins. It is also clear that the plate number is of little

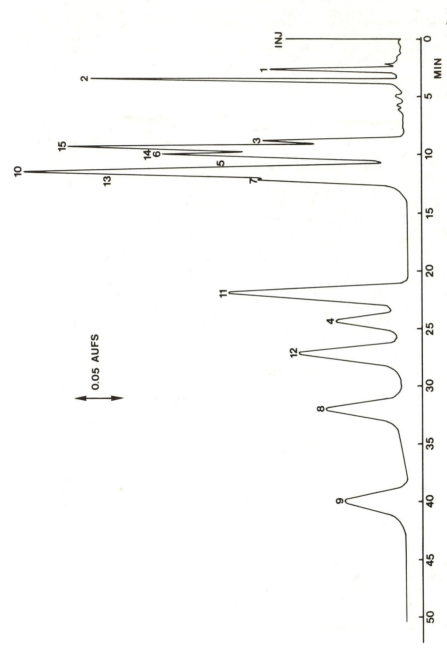

Figure 4. Chromatogram obtained on LiChrosorb RP-8 10 um Hibar with mobile phase acetonitrile-water-0.2 M phosphate buffer pH 7.0 (8.5:86.5:5).

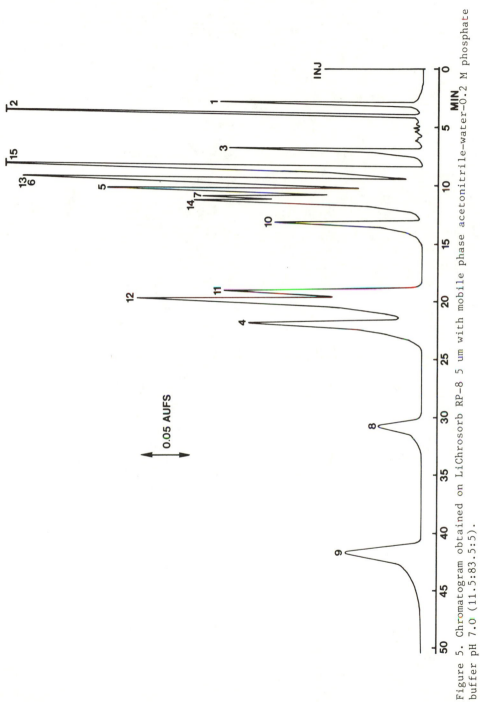

Figure 5. Chromatogram obtained on LiChrosorb RP-8 5 um with mobile phase acetonitrile–water–0.2 M phosphate buffer pH 7.0 (11.5:83.5:5).

Table II : ORDER OF ELUTION OF CEPHALOSPORINS

Column	Order of elution of cephalosporins
LiChrosorb RP-8 10 μm Hibar	1 2 3 15 6 14 5 [10] 13 (7) 11 4 12 8 9
LiChrosorb RP-8 10 μm	1 2 15 14 3 6 5 [10] 13 (7) 4 12 11 8 9
LiChrosorb RP-8 5 μm	1 2 3 15 6 13 14 5 (7) [10] 11 12 4 8 9
Zorbax C8 7 μm	1 2 [10] 5 6 14 3 15 (7) 11 13 4 8 9 12
Polygosil C8 10 μm	1 2 [10] 14 15 5 6 13 3 (7) 8 9 4 12 11
Nucleosil C8 10 μm	1 2 [10] 14 15 5 6 3 13 (7) 4 8 9 12 11
RSil C18 LL 10 μm	1 2 [10] 14 15 5 6 3 13 (7) 8 9 4 12 11
μBondapak C18 10 μm	1 2 15 3 6 14 5 [10] (7) 13 12 4 11 8 9
Nucleosil C18 10 μm	1 2 15 14 [10] 3 5 6 13 (7) 4 12 8 9 11
Partisil ODS 10 μm	1 2 [10] 14 15 5 6 3 13 (7) 8 9 4 12 11

value to estimate the usefulness of a column for a particular separation problem.

In Table III other parameters generally provided by manufacturers are listed. In the second column the specific surface area of the parent silica as stated by the manufacturer is given. In the third column our results from BET measurements on parent silicas of the same type are given. For LiChrosorb 5 µm, the BET measurement was 365, while the catalogue value was 500. For Zorbax there is only a small difference between the value of 289 we obtained and the catalogue value of 300. For Polygosil we do not find half of the value stated by the manufacturer. On the other hand, for Nucleosil, from the same manufacturer, the obtained value of 347 agrees quite well with the catalogue value of 300. For Partisil good agreement is observed and less so for RSil.

Pore diameters stated by the manufacturer are also listed. As can be seen, it is not common to use wide pores for longer chain lengths. We also determined the mean diameter by measuring the total pore volume of parent silica of the same type by means of water titration (11). The mean pore diameter is calculated from the measured BET-values and pore volumes and the pores are assumed to be cylindrical. For pore diameters, there are also large differences between nominal values and measured values. For Zorbax and RSil, the nominal value agrees with the measured value of 8 nm. For Polygosil a large difference is observed ; we measured 16 nm, the claimed value is 6 nm.

Carbon content values are taken from the literature. They are expressed as percentages of the total mass. We did not determine carbon content, but the loss of mass on ignition at 700 °C was determined. For the loss on ignition one would expect to find higher figures than for the carbon content, since the former includes the loss of hydrogen and some water from the silica and also since it is expressed here versus the mass of the residue obtained after ignition, which corresponds approximately to the amount of bare silica involved in the preparation of the reversed phase. Indeed, when a sample of silica is heated under the same conditions, a loss on ignition of about 3 % is obtained. The carbon content is expressed versus the mass of derivatized material. It is also recommended to express carbon content and loss on ignition as a function of the specific surface area of the parent silica which is supposed to give a better idea of the loading per unit surface. The values between brackets are normalized for a surface area of 100 m^2 starting from our own BET measurements. For several columns, loss on ignition and carbon content are quite close; for others, especially Zorbax, a striking difference is observed.

Very often it is accepted that the higher the loading, the higher the acetonitrile content should be in the mobile phase to obtain elution. It is observed that the percentages needed in order to elute the cephalosporins within 40 min are very different and no clear correlation can be seen with the carbon content. However, not only the amount of organic material plays a role, but also the residual silanol activity. Therefore we modified the methyl red adsorption test of Shapiro and Kolthoff (13) in order to use it for the quantitative measurement of residual silanol groups. A stock solution of 400 mg/100 ml of methyl red was used and the results are given in Table III. For Zorbax, a very low methyl red adsorption

Table III : CHARACTERISTICS OF PACKING MATERIALS AND COLUMNS

Packing Materials	Surface area of the parent silica (m²/g)		Pore diameter of parent silica (nm) (a)	Mean pore size of parent silica (nm)	Carbon content (% m/m) (b,e)	Loss on ignition (% m/m) (c,e)	CH_3CN in mobile phase (% V/V)	Methyl red adsorption value (mg/g) (e)	Capacity factor XII
	Claimed (a)	Results (d)							
LiChrosorb RP-8 10 μm Hibar (f)	500	NA	6	NA	13-14	ND	8.5	ND	9.8
LiChrosorb RP-8 10 μm	500	NA	6	NA	13-14	14.5	8.5	2	11.9
LiChrosorb RP-8 5 μm	500	365	6	12	13-14	16.7 (45.8)	11.5	2 (0.5)	5.8
Zorbax C8 7 μm	300	289	6-8	8	15	7.6 (26.2)	1	0.5 (0.2)	2.8
Polygosil C8 10 μm	500	234	6	16	10-11	9.8 (41.8)	5	15 (6.4)	18.4
Nucleosil C8 10 μm	300	347	10	13	10-11	10.6 (30.6)	5	13 (3.7)	18.9
RSil C18 LL 10 μm	550	439	6	8	10	17.9 (40.8)	10	92 (21.0)	27.7
μBondapak C18 10 μm	350	NA	10	NA	10	11.5	10	1	8.6
Nucleosil C18 10 μm (f)	300	347	10	13	15-16	ND	7	ND	15.3
Partisil ODS 10 μm (f)	360	394	6	9	5	ND	5	ND	23.3

a: manufacturer's value, b: values taken from literature [12], c: 4 hours at 700 °C, d: results from BET measurements obtained on parent silica of the same type, e: figures between brackets are calculated using own BET values and are expressed in mg/100 m², f: prepacked columns, ND: not determined since columns were prepacked, NA: not available

value of 0.5 is obtained which means that it is well covered material. The high methyl red adsorption value for RSil, i.e. 92, corresponds well with the labelled low loading (LL), but is contradic-
tory to the fact that this material is said to be endcapped to ensure minimum silanol activity. The methyl red adsorption value corresponds quite well with capacity factors of cefaloridine carrying a
positive charge on the pyridinium ring. The high methyl red adsorption value of 92 for RSil corresponds with a capacity factor of 27
for cefaloridine. It indicates that chromatography of cefaloridine
could be used to check for free silanol groups on the packed column.

Conclusions

It can be concluded that chromatographic materials classified under
the general label of C8 or C18, can behave in very different ways
providing a wide range of unique selectivities. Not all the manufacturers are able to produce chromatographic materials with sufficient
reproducibility over longer periods. A particular separation cannot
always be reproduced on another column of the same type or from the
same manufacturer. The information given by the manufacturers, such
as the plate number obtained in a test chromatogram, is often in-
sufficient or irrelevant to evaluate column performance. Other parameters neither give information pertaining to selectivity. The parameters discussed in this chapter have been found useful to compare
different reversed phase materials with each other and to check the
quality of different batches of the same material. Great care is
recommended when a high performance liquid chromatographic method is
planned for application in several laboratories. The proposed chromatographic system should be tried with a series of different
packing materials of the same type to check for possible influence
of the packing material on the selectivity. If no significant influence is observed, as for example for the phenanthrene–anthracene
separation mentioned above, the simple indication of the column type
C8 or C18 may be sufficient. But if there is an influence, as is the
case with the cephalosporins, a selectivity test should be provided.
This necessitates the use of reference materials which will render
the method less attractive in cases where the reference products are
not commonly available.
    Citation in published methods of the manufacturing brands of
packing materials known to give good results is required for docu-
mentation, it is not justifiable on scientific grounds as long as
manufacturers do not prepare packing materials with better reproducibility.

Literature Cited

1. Nice, E.C.; O'Hare, M.J. J. Chromatogr. 1978, 166, 263.
2. Ogan, K.; Katz, E. J. Chromatogr. 1980, 188, 115.
3. Colmsjö, A.L.; MacDonald, J.C. Chromatographia 1980, 13, 350.
4. Panthananickal, A.; Marnett, L.J. J. Chromatogr. 1981, 206, 253.
5. DiCesare, J.L.; Dong, M.W. Perkin-Elmer Corporation,
   Chromatography Newsletter 1982, 10, 12.
6. Engelhardt, H.; Dreyer, B.; Schmidt, H. Chromatographia 1982,
   16, 11.
7. Goldberg, A.P. Anal. Chem. 1982, 54, 342.

8. Shaikh, B.; Tomaszewski, J.E. Chromatographia 1983, 17, 675.
9. Landy, J.S.; Ward, J.L.; Dorsey, J.G. J. Chromatogr.Sci. 1983, 21, 49.
10. Wouters, I.; Hendrickx, S.; Roets, E.; Hoogmartens, J.; Vanderhaeghe, H. J. Chromatogr. 1984, 291, 59.
11. Verzele, M.; Lammens, J.; Van Roelenbosch, M. J. Chromatogr. 1979, 186, 435.
12. Majors, R.E. J. Chromatogr. Sci., 1980, 18, 488.
13. Wouters, I.; Quintens, I; Roets, E.; Hoogmartens, J. J. Liquid Chromatogr. 1982, 5, 25.

RECEIVED October 10, 1985

Selectivity of Poly(styrene–divinylbenzene) Columns

John M. Joseph

Squibb Institute for Medical Research, New Brunswick, NJ 08903

Macroporous poly(styrene-divinylbenzene) copolymer,
PRP-1, columns were used as the stationary phase in the
reverse-phase HPLC of the synthetic β-lactam antibiotic
aztreonam [2S-[2α,3β(Z)]]-3-[[(2-Amino-4-thiazolyl)[(1-
carboxy-1-methylethoxy)-imino]acetyl]amino]-2-methyl-4-
oxo-1-azetidinesulfonic acid and related compounds.
Aztreonam was separated better from its precursors and
therefore could be assayed more accurately. In most
cases, the elution order of compounds tested on a PRP-1
column followed that in conventional reverse-phase,
suggesting a similar separation mechanism. For various
separations investigated, PRP-1 was found to be more
suitable for our applications than the silica-based
reversed-phase columns.

Reverse-phase packings of octylsilane (C_8) and octadecylsilane (C_{18})
are by far the most popular and widely used stationary phases in
modern liquid chromatography. In spite of their wide popularity as
universal liquid chromatography phases, these alkyl bonded phases
(silica packings, in general) have several shortcomings, the major
one being their instability in aqueous phases (1,2). The possibilities for the regulation of retention and selectivity by ion
suppression and complexation are limited in the reverse-phase mode
with functionalized silica, as degradation of the stationary phase
occurs outside the pH range 2-8 (3). Column-to-column variation is
another major disadvantage with bonded phase columns. Significant
variations in selectivity and retention characteristics exist among
bonded phase columns from manufacturer to manufacturer, and even
within lot to lot from the same manufacturer. Several reasons are
attributed to this problem. A few are worth mentioning as they inherently are associated with silica-based supports. Due to steric
hindrance, the bonding reaction is always incomplete leaving unreacted silanol groups (Si-OH) on the silica surface (4). These
residual silanol groups can cause mixed retention mechanisms,
especially with basic compounds, leading to non-reproducibility in
retention from batch-to-batch of columns (5). In addition, residual

Si-OH groups are also focal points of attack for water and other reagents in the mobile phase, causing dissolution of underlying silica with gradual deterioration in column performance and eventual reduction in column lifetime (5,6). The pH of the silica surface also varies from acidic, neutral, to basic, and is believed to influence the preparation and properties of bonded phases (7). The inclusion of traces of transition metals in silica matrix can further influence the retention of acids, bases or neutral compounds that can undergo complexation reactions (1). Thus the selectivity and retention characteristics of bonded phase columns also depend on the quality of silica used for the bonding reactions.

The use of macroporous poly(styrene-divinylbenzene) copolymer, PRP-1, as reverse-phase adsorbent has been previously reported in the analysis of pharmaceuticals (8,9), organic acids (8), nucleosides (10), plant hormones (11), and phenolic compounds (12). The major advantage of this resin-based packing material is its extreme stability over a wide pH range of 1-13 (10). The chemical stability of PRP-1 at elevated pH levels has enabled new applications. The increased detectability for barbiturates (13) and thiamine derivatives (14) was reported using PRP-1 column with mobile phase, at elevated pH levels (pH 8-9). Similarly, PRP-1 functionalized with trimethyl amine (PRP-X100, anion-exchanger) has been used with basic eluents in ion-chromatography using conductivity detection (15).

This paper describes the use of poly(styrene-divinylbenzene) copolymer, PRP-1, as a reverse-phase adsorbent in the assay of the antibiotic aztreonam and related compounds. Comparisons are also made for similar assays using silica-based columns. None of the shortcomings described earlier, associated with bonded phase columns, is observed. In addition to the reverse-phase mode, the PRP-1 columns are tested in ion-pair as well as in size exclusion modes of separation. Superior resolutions are obtained in the reverse-phase chromatography of ionic compounds without the use of ion-pairing agents. In addition to the normal adsorption and/or partitioning, $\pi-\pi$ interaction is also seen to play a dominant role in the separation of aromatic compounds using a PRP-1 column.

Experimental

The modular HPLC system used in this study included the following components: a Model 110A Altex pump (Beckman Scientific Instruments) or a System 4 Series programmable pump (Perkin-Elmer Corporation), an in-line filter, a Model 773 variable-wavelength UV detector (Kratos Scientific Instruments) fitted with a 12 μL flow cell, a Model 600 Series autosampler with a nominal volume of 20 μL (Perkin Elmer Corporation) and a two channel recorder (Kipp and Zonen). The chromatographic data were processed with the aid of a Model 3357 laboratory computer (Hewlett-Packard).

The chromatographic separations were carried out either on a 15 or 25 cm x 4.6 mm I.D. stainless-steel column packed with 10 μm poly(styrene-divinylbenzene) copolymer (PRP-1, Hamilton). The conventional reverse-phase chromatography was performed using either a 10 μm Whatman Partisil ODS-3 (25 cm x 4.6 mm I.D.) or a Waters 10 μm μBondapak C_{18} (30 cm x 3.9 mm I.D.) column. The analytical columns were kept at 31° using a column heater (Bioanalytical Systems, Inc.,

LC-23A) unless otherwise specified. Since PRP-1 is a non-silaceous particle, no precolumn or saturator column was used. The silica-based analytical columns were used in conjunction with a precolumn packed with 37 μm silica.

The solvents used were: (a) HPLC grade methanol and acetonitrile (Fisher Scientific) (b) water, double-distilled and stored in glass. Inorganic salts and acid (ammonium sulfate, potassium phosphate monobasic, potassium phosphate dibasic and phosphoric acid, 85%) were analytical reagent grade (Mallinckrodt Chemical Works).

The mobile phase was delivered at a flow rate of 1-1.5 mL/min. The pH of the mobile phase was adjusted before the addition of organic modifiers.

Results and Discussion

Aztreonam is a β-lactam antibiotic with a wide spectrum of activity against a variety of gram-negative organisms including Pseudomonas aeruginosa (16). The synthesis of this compound starting from the amino acid, L-threonine has been previously reported (17). Figure 1 shows the structure of aztreonam and related compounds.

Aztreonam, the (Z)-isomer, is the biologically active form. The alternative configuration, the (E)-isomer, was synthesized. The open ring form (Z) obtained by the cleavage of the β-lactam moiety is the principal biological metabolite. Other possible, related compounds include the ethyl ester of aztreonam, which could be formed in trace amounts during recrystallization of aztreonam from ethanol, desulfonated aztreonam, the open-ring form (E) and the desulfonated open-ring form.

Several HPLC systems capable of resolving aztreonam from possible related compounds, mainly the (E)-isomer and the open ring form (Z) were examined. As shown in Figure 2, the conventional reverse-phase system utilizing a Waters μBondapak C_{18} column was found to be suitable for carrying out these separations. The mobile phase, here, is composed of 82% phosphate buffer (0.05M, pH 3.0) and 18% methanol. The desulfonated aztreonam almost coeluted with the open-ring form (Z) in this chromatographic system. As expected, the aztreonam ethyl ester was retained longer (RT ∿ 24 minutes) than the rest of the compounds due to its increased hydrophobicity.

An inherent problem and perhaps, one of the biggest disadvantages of silica-based columns, in general, is the column-to-column variation (1). This was clearly manifested during an attempt to reproduce the separation profile obtained on a Waters μBondapak C_{18} column using other conventional C_{18} columns from diverse manufacturers (Whatman, E. Merck, IBM, etc.) under similar and other chromatographic conditions. Even though the separation of the open-ring form (Z) and the ethyl ester from aztreonam was accomplished, these reverse-phase columns lacked the selectivity to effect adequate resolution of the (Z)- and (E)- isomers of aztreonam. Similarly, amine columns (silica bonded to n-propyl amine) from various manufacturers (IBM, Whatman, DuPont, and Waters) were also examined in the same capacity using a mobile phase of methanolic phosphate buffer (pH 3.5) containing small amounts of tetrabutylammonium hydrogen sulfate (18). Here also, the major problem was column-to-column variation, which once again, proved to be real rather than

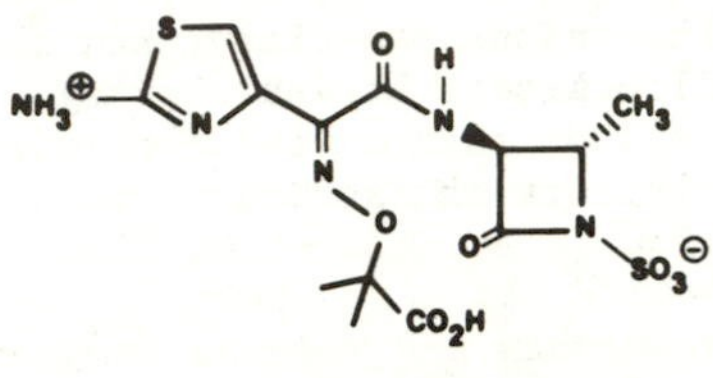

AZTREONAM (Z)

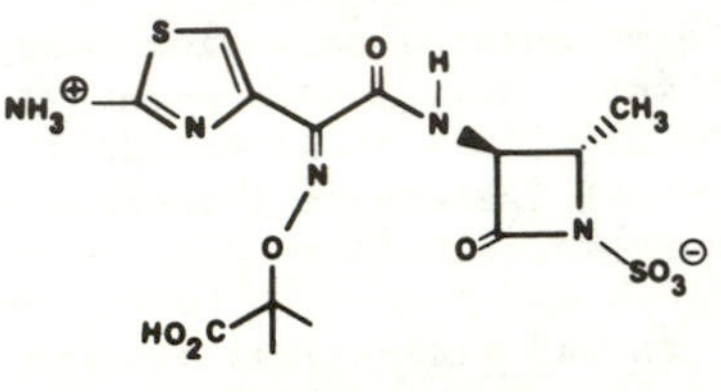

AZTREONAM (E)

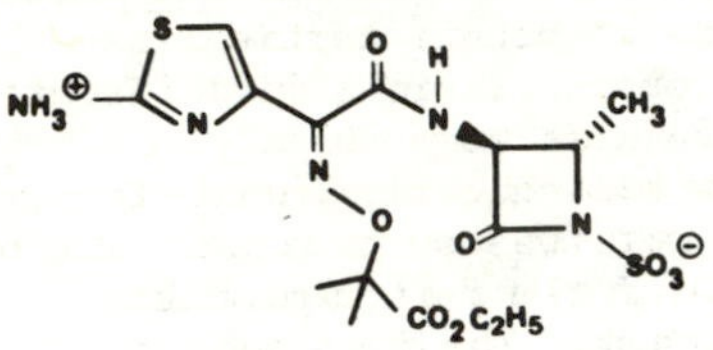

AZTREONAM ETHYL ESTER

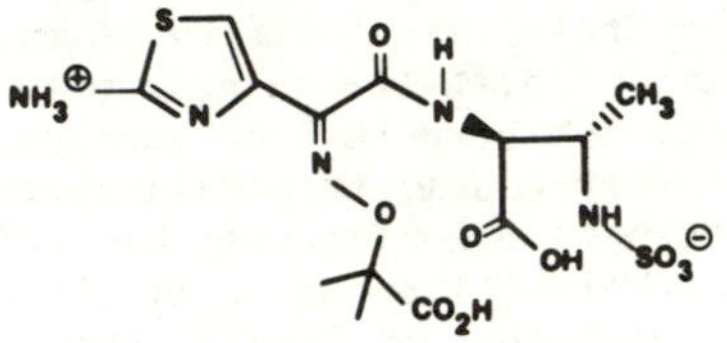

OPEN-RING FORM (Z)

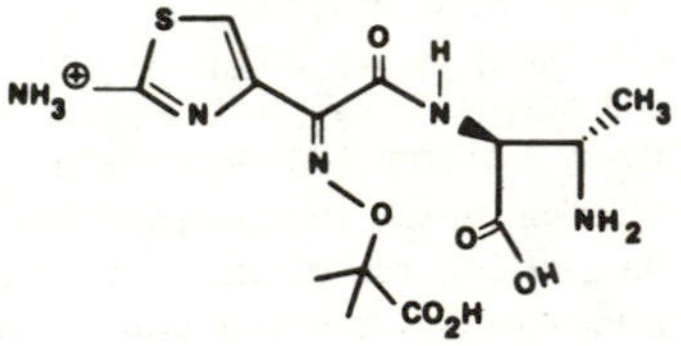

DESULFONATED OPEN-RING FORM

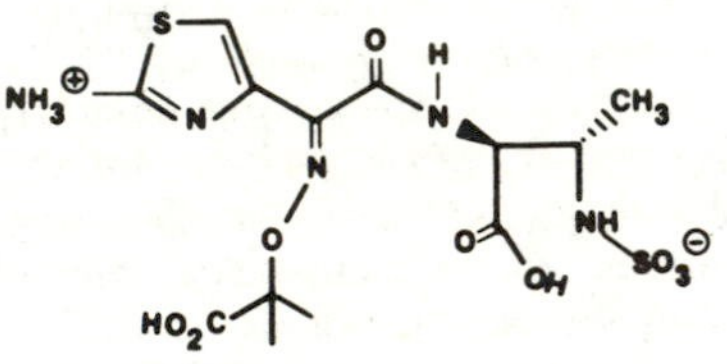

OPEN-RING FORM (E)

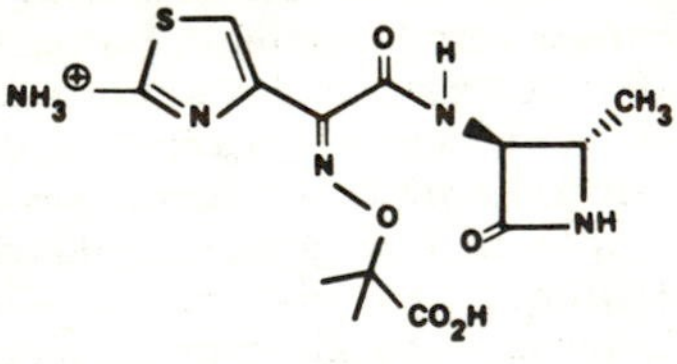

DESULFONATED AZTREONAM

Figure 1. Structure of aztreonam and related compounds.

apparent with silica based columns. Only the Zorbax amine column
(DuPont) provided adequate separation of aztreonam from related
compounds including the (E)-isomer. In this instance, the mode of
separation presumably is ion-exchange (weak anion-exchanger). The
elution order of aztreonam ethyl ester and open-ring form (Z) were
found to be the reverse of that found in conventional reverse-phase
HPLC (Figure 3).

The less polar aztreonam ethyl ester was eluted first, and the
relatively more polar open-ring form (Z) became the last eluting
component in this chromatographic system. No change was observed in
the elution order of the (Z)- and (E)- isomers of aztreonam. One
difficulty with an amine column is the longer equilibration time
required as opposed to reverse-phase (bonded phase and PRP-1).

Figure 4 shows the separation of aztreonam and related compounds
accomplished on a PRP-1 column. The mobile phase was essentially the
same as that used in conventional reverse-phase except that a small
amount of acetonitrile (4%) was incorporated to sharpen the peaks.
Selectivity was found to be much better with this resin-based
stationary phase compared to a bonded phase, as it afforded superior
and otherwise difficult separations of these very closely related
compounds. The improved resolution between the desulfonated open-
ring form and open-ring form (Z) would enable accurate quantitation
of both of these compounds if they were present as impurities in
bulk material or formulations of aztreonam. The most noteworthy was
the excellent resolution obtained between the (Z)- and (E)- isomers
of the open ring form of aztreonam. It was found that the (Z)-
isomer always eluted before the corresponding (E)- isomer in both
C_{18} and PRP-1 columns (Figures 2 and 4). Based on this, it was
presumed that the peak at RT 3.9 minutes (Figure 4) might be the
(E)-isomer of open-ring desulfonated aztreonam. The aztreonam
ethyl ester was retained longer on a PRP-1 column (RT $\sim$ 35 min.) as
opposed to conventional C_{18} columns. The polystyrene matrix is
believed to be extremely hydrophobic (6), which accounts for the
prolonged retention of relatively non-polar entities like aztreonam
ethyl ester. Except desulfonated aztreonam which was unavailable at
the time of this experiment, the elution order of all the other
compounds from a PRP-1 column followed that in conventional reverse-
phase, suggesting a somewhat similar retention mechanism under
identical chromatographic conditions. The desulfonated aztreonam was
retained longer ($\sim$ 16 minutes), and it was eluted after the 'E'-
isomer. Another type of interaction which is unique to PRP-1 is
π-π interaction, especially with aromatic solutes (19). Figure 5
shows the separation obtained on a 15-cm PRP-1 column under the same
chromatographic conditions. No loss of resolution was noted. The
retention time was lowered with shorter columns. Concomitantly, the
peaks appeared to be much sharper.

The PRP-1 column also has been used successfully in the HPLC
assay of several key aztreonam intermediates. The carbobenzoxy-L-
threoninamide (CBZ-L-threoninamide) was one. It was the first major
intermediate isolated in the synthesis of aztreonam starting from
L-threonine. The synthetic sequence is depicted in Figure 6. Two
major impurities were benzyl alcohol and benzyl carbamate; the former
produced by the hydrolysis of benzyl chloroformate and the latter by
the reaction between benzyl chloroformate and ammonia. The purpose

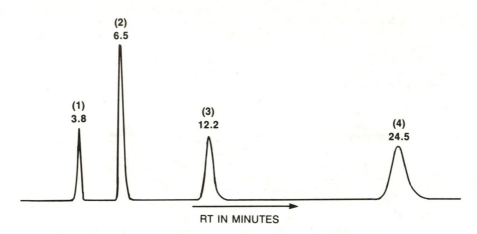

Figure 2. Conventional reverse-phase HPLC of aztreonam and related compounds. Column: Waters μbondapak C_{18} (30 x 3.9 mm i.d.).
Mobile phase: 82% 0.05$\underline{M}$ phosphate buffer (pH 3.0)-18% methanol.
Flow rate: 1.5 mL/min. Detection: UV 254 nm. Peaks: (1) = open-
ring form (Z), (2) = aztreonam (Z), (3) = aztreonam (E), (4)=
aztreonam ethyl ester.

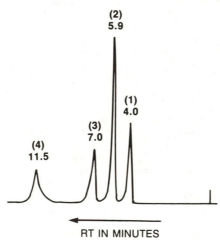

Figure 3. Separation of aztreonam and related compounds using a
Zorbax amine column (25 x 4.6 mm i.d., 30°C). Mobile phase: 97.5%
0.1$\underline{M}$ phosphate buffer-5 m$\underline{M}$ TBAHS (pH 3.5) - 2.5% methanol. Flow
rate: 1 mL/min. Detection: UV 220 nm. Peaks: (1) = aztreonam
ethyl ester, (2) = aztreonam (Z), (3) = aztreonam (E), (4) =
open-ring form (Z).

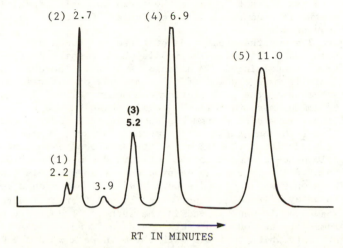

(2) 2.7   (4) 6.9

(5) 11.0

(3)
5.2

(1)
2.2

3.9

RT IN MINUTES

Figure 4. Separation of aztrenam and related compounds using a PRP–1 column (25 x
4.1 mm i.d., 31 C). Mobile phase: 83% 0.05 M phosphate buffer (pH 3.0)– 13% methanol –
4% acetonitrile. Flow rate: 1 mL/min. Detection: UV 254 nm. Peaks: (1) = open–ring
desulfonated form, (2) = open–ring form (Z), (3) = open–ring form (E), (4) =
aztreonam (Z), (5) = aztreonam (E).

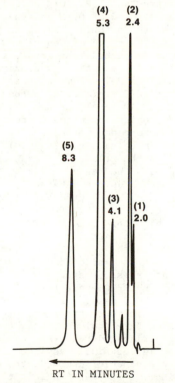

(4)   (2)
5.3   2.4

(5)
8.3

(3)   (1)
4.1   2.0

RT IN MINUTES

Figure 5. Separation of aztreonam and related compounds on a 15–cm PRP–1 column.
Chromatographic conditions and elution order are the same as in Figure 4.

of the HPLC system was to resolve these two possible impurities and
the two synthetic precursors, L-threonine and CBZ-L-threonine, from
CBZ-L-threoninamide.

Figure 7 shows the separation of these compounds on a 15-cm
(i.d. 4.6 mm) PRP-1 column. The mobile-phase consisted of methanol-
water (60:40) containing 0.05M ammonium sulfate. The chromatography
provided excellent resolution of all compounds of interest with
benzyl carbamate as the last eluting component (RT ∿ 8.9 minutes).
The use of ammonium sulfate in the mobile phase helped to reduce
peak tailing. This HPLC system has now been used for the routine
assay of CBZ-L-threoninamide. This method is fast, rugged, sensitive and reproducible. Column stability was found excellent with
minimum column-to-column variation. Figure 8 shows the chromatogram
of a batch of CBZ-L-threoninamide produced by an overseas manufacturer. The sample was assigned a purity of 100% based on a reverse-
phase HPLC assay (C_{18} column) performed by a different laboratory.
When the same sample was assayed in the author's laboratory using a
PRP-1 column, it was found to contain a significant amount (∿ 5%) of
benzyl carbamate which coeluted with the main peak when the ODS
column was used. Consequently, the laboratory using the common ODS
column showed erroneously high purity results.

In the meantime, attempts to develop a reverse-phase HPLC assay
for CBZ-L-threoninamide were vigorously being made in other laboratories. Eventually, a reverse-phase HPLC assay was developed for
this compound using a Waters Radial Pak C_{18} cartridge (10 cm x 8 mm
i.d.) and a mobile phase composed of 0.1M phosphate buffer-methanol
(82:18). Figure 9 represents the chromatographic profile. Even
though the most difficult separation of benzyl carbamate from
CBZ-L-threoninamide was accomplished, the greatest shortcoming of
this method was the prolonged assay time: approximately 55 minutes,
six times longer than when a PRP-1 column was used. Such a lengthy
assay time is less attractive for practical purposes.

Styrene-divinylbenzene XAD copolymer was shown to give higher
K' values for amino acids containing aromatic residues (e.g.,phenyl
alanine, tyrosine, etc.) as opposed to those with simple aliphatic
moieties; for example, valine, leucine, etc. (20). This increased
retention was attributed to the greater interaction of the aromatic
moiety of the amino acid with the resin matrix by π-π interaction
(20). With an ODS column such unique interaction (π-π) is non-
existent. The ability of the phenyl group of benzyl carbamate to
exert this type of interaction with the PRP-1 matrix might have
caused its easy separation on this column (Figure 10).

Other efficient separations accomplished using a PRP-1 column
were of compounds involved in the synthesis of carbobenzoxy-L-serin-
amide mesylate starting from the amino acid, L-serine (Figure 11).
The purpose of the chromatographic system was to aid organic chemists
at various synthetic steps by HPLC monitoring of the reaction intermediates. Fast and simple HPLC assays are crucial to this type of
work, and in most cases, as in the current example, efficient HPLC
assays were developed using PRP-1 columns. All three synthetic precursors (L-serine, CBZ-L-serine and CBZ-L-serinamide) were presumed
to be possible impurities in the final mesylate intermediate. As
shown in Figure 12, these synthetic intermediates were completely
resolved from CBZ-L-serinamide mesylate in less than eight minutes.

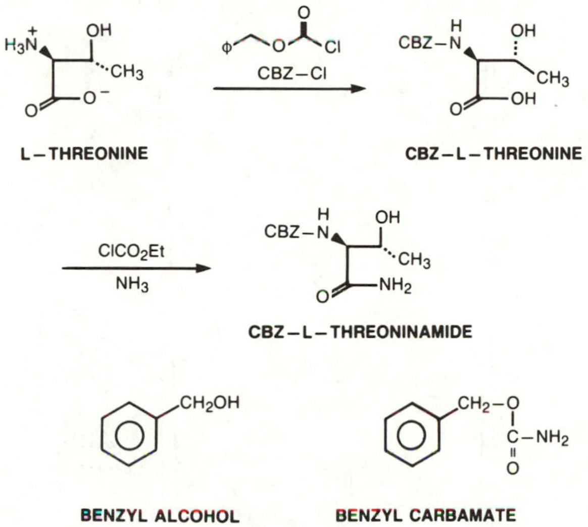

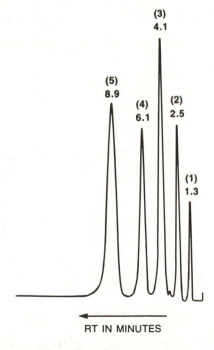

Figure 6.   Synthesis of CBZ–L–threoninamide.

Figure 7.   HPLC of CBZ–L–threoninamide. Column: PRP-1 (15 x 4.1 mm i.d., 30°C). Mobile phase: 60% methanol – 40% water – 0.05M ammonium sulfate. Flow rate: 1 mL/min. Detection: UV 215 nm. Peaks: (1) = L-threonine, (2) = CBZ–L–threonine, (3) = CBZ–L–threoninamide, (4) = benzyl alcohol, (5) = benzyl carbamate.

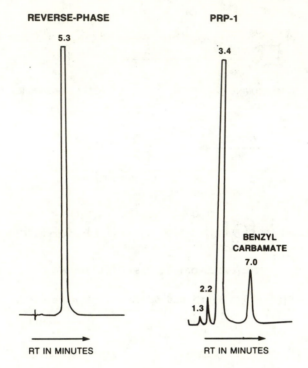

Figure 8.   Chromatograms of a batch of CBZ–L-threoninamide using C₁₈ and PRP-1 columns.

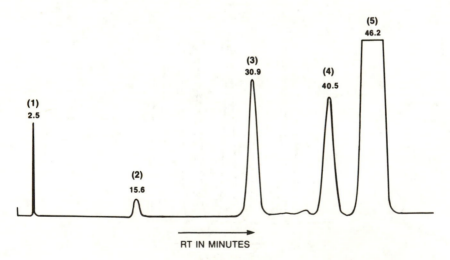

Figure 9.   HPLC of CBZ–L-threoninamide using a Waters Radial Pak C₁₈ Cartridge (10 x 8 mm i.d.).   Mobile phase: 82% 0.1M phosphate buffer (pH 3.0) - 18% methanol.   Flow rate: 2 mL/min. Detection: UV 204 nm.   Peaks: (1) = L-threonine, (2) = benzyl alcohol, (3)= CBZ–L-threonine, (4) = benzyl carbamate, (5) = CBZ–L-threonina-mide.

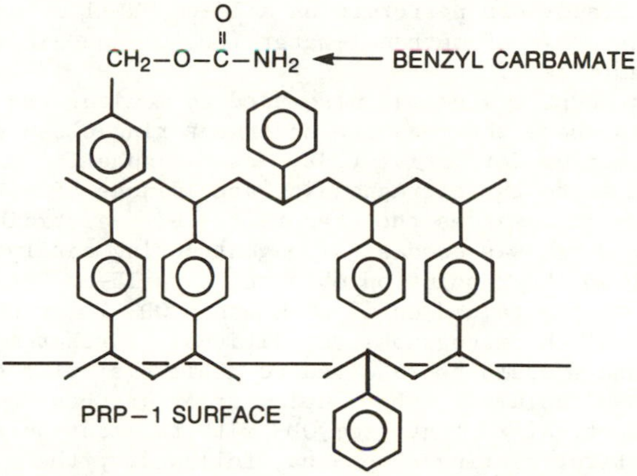

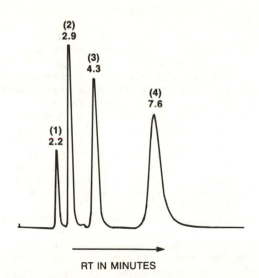

$$O$$
$$\|$$
$$CH_2-O-C-NH_2 \longleftarrow \text{BENZYL CARBAMATE}$$

PRP–1 SURFACE

Figure 10. Hypothetical π–π interaction of phenyl group of benzyl carbamate with PRP-1 matrix.

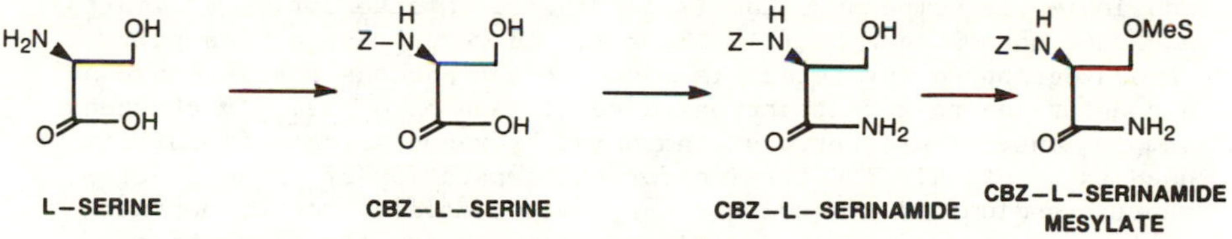

L–SERINE   CBZ–L–SERINE   CBZ–L–SERINAMIDE   CBZ–L–SERINAMIDE MESYLATE

Figure 11. Synthesis of CBZ-L-serinamide mesylate.

(2)
2.9

(3)
4.3

(4)
7.6

(1)
2.2

RT IN MINUTES

Figure 12. HPLC of CBZ-L-serinamide mesylate. Column: PRP-1 (15 x 4.1 mm i.d., 30°C). Mobile phase: 60% methanol – 40% water – ammonium sulfate (1 gm/liter). Flow rate: 1 mL/minute. Detection: UV 210 nm. Peaks: (1) = L-serine, (2) = CBZ-L-serine, (3) = CBZ-L-serinamide, (4) = CBZ-L-serinamide mesylate.

The chromatography was performed on a 15-cm PRP-1 column with a simple mobile phase of methanol-water (60:40) containing ammonium sulfate (1 gm/liter).

The same HPLC system was also used to monitor the next step of the synthesis where the mesylate underwent ring closure to form its tetrabutylammonium derivative (CBZ-L-azetidinone-TBA salt) which was finally reduced to the corresponding azetidinone (Figure 13). Another possible impurity besides the precursor, CBZ-L-azetidinone-TBA salt, was lutidine which was used as a reagent during catalytic reduction. Figure 14 shows the resolution obtained on a 15-cm PRP-1 column as compared to that obtained on 25-cm Whatman ODS-1 and ODS-3 columns under identical chromatographic conditions. Further modification of the mobile phase would be required to achieve similar separations on ODS-1 and ODS-3 columns. The elution order of these compounds from PRP-1 was identical to that from ODS with the most polar TBA derivative as the first eluting component, followed by the less polar CBZ-L-azetidinone. Lutidine being least polar among the three was the last to elute from all three columns.

Ion-pair chromatography has dramatically extended the scope of bonded reverse-phase HPLC, as it successfully separates both ionic and ionizable compounds, and is considered an alternative to ion-exchange chromatography (21, 22, 23). In reverse-phase ion-pair chromatography a surfactant is added to the aqueous mobile phase as a counter-ion to effect increased resolution of oppositely charged sample ions. Thus, tetrabutylammonium hydrogen sulfate (TBAHS) is used as a cationic counter-ion for the separation of organic acids, whereas sodium dodecyl sulfate (or, more commonly, sodium heptane sulfonate) is used as an anionic counter-ion for the separation of organic bases. Various theories have been proposed to explain the mechanism of the ion-pair phenomenon. A complete discussion of the mechanistic aspects of ion-pair chromatography has been given by Bidlingmeyer (22).

The objective, here, was to examine the retention characteristics of a PRP-1 column as opposed to ODS columns in ion-pair chromatography using TBAHS as the counter-ion. The compound I (Figure 15) is CBZ-L-azetidinone-TBA salt derived from L-threonine, and it is another important intermediate in the synthesis of aztreonam. Compounds II and III are possible impurities associated with I. A reverse-phase ion-pair chromatographic method was originally developed for I using a Whatman ODS-1 column and a mobile phase composed of methanol-water (70:30) containing 0.1M ammonium sulfate and 5 mM TBAHS (pH 5.0). Figure 16 shows the separation obtained for I, II and III. The chromatography took ∿ 11 minutes with CBZ-L-azetidinione-TBA salt (compound I) as the last eluting component.

When the separation was carried out under similar conditions using a 25-cm PRP-1 column, no peak was visible even after a period of ∿ 50 minutes. All three test compounds were believed to be strongly retained on the PRP-1 column when TBAHS was used as the counter-ion. A fresh batch of the same mobile phase was prepared without TBAHS and the chromatography was repeated with another 25-cm PRP-1 column. The test compounds eluted this time with wide separation in less than 30 minutes (Figure 17). On the contrary, loss of resolution for the test compounds I, II and III was observed when the chromatography was repeated using the same Whatman ODS-1 column with

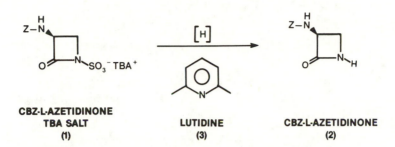

Figure 13.   Synthesis of CBZ-L-aztetidinone.

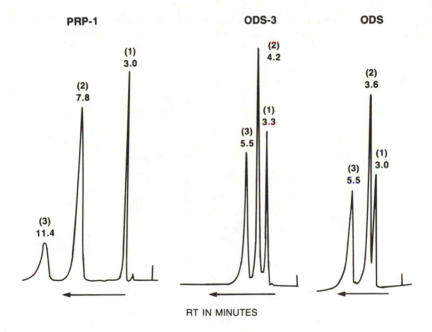

RT IN MINUTES

Figure 14.   Comparison of separations on PRP-1 vs C₁₈ columns.
Mobile phase: 60% methanol - 40% water - ammonium sulfate (1 gm/
liter). Flow rate: 1 mL/min. Detection: UV 215 nm. Peaks: (1) =
CBZ-L-azetidinone- TBA salt, (2) = CBZ-L-azetidinone, (3) =
lutidine.

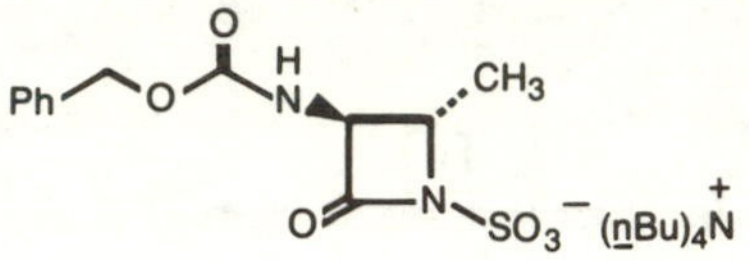

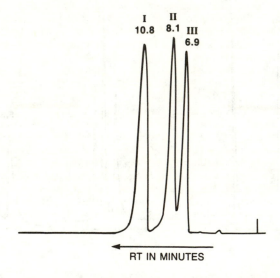

Figure 15.   Structure of CBZ-L-azetidinone-TBA salt and related compounds derived from L-threonine.

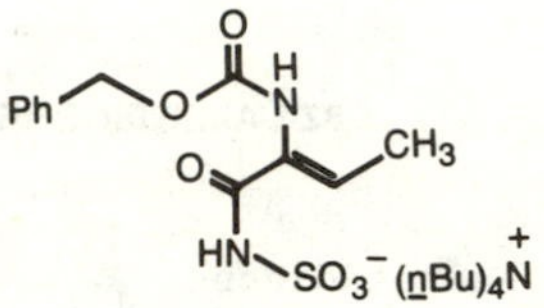

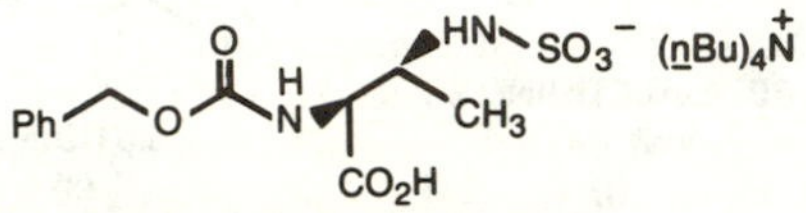

Figure 16.   HPLC of CBZ-L-azetidinone- TBA salt derived from L-threonine using bonded reverse-phase ion-pair chromatography. Column: C_{18} (25 x 4.6 mm i.d., 30°C). Mobile phase: 70% water - 5 mM TBAHS (pH 5.0) - 30% methanol - 0.1M ammonium sulfate. Flow rate: 1 mL/min. Detection: UV 215 nm.

the modified mobile phase without TBAHS. Evidently, TBAHS helped to enhance the retention of ionic species on a bonded reverse-phase, thereby effecting their separation. Unlike C_{18}, poly(styrene-divinylbenzene) matrix, by virtue of its polymer structure, possesses strong adsorbent properties (24). It is now evident that PRP-1 could retain moderately ionic species (here, anions) better than bonded reverse-phase columns without further modifications of its polymer surface with large organic counter-ions like TBAHS.

A small amount of TBAHS could be used to improve separation in the case of fairly ionic compounds which show poor retention on PRP-1 columns. As shown in Figure 18, the marginal resolution obtained between aminoxyisobutyric acid and phthalic acid was improved by adding 1 m$\underline{M}$ TBAHS to the same mobile phase of acetonitrile-water (10:90) containing ammonium sulfate (4 gm/liter). TBAHS seemed to have no effect on the retention of aminoxyisobutyric acid, but it enhanced the retention of phthalic acid considerably (from 2.9 minutes to 25.9 minutes).

Exclusion effects are also possible with PRP-1 column depending on the size of the solute molecule. This was evident in the retention behavior of the macrocyclic heptaene antibiotic, amphotericin B, on a PRP-1 column. The reverse-phase fast LC method recently developed in the author's laboratory using a 5-cm long C_{18} column (3 micron particle size) gave a retention time of 5.5 minutes for this compound. This method also provided excellent resolution of amphotericin B from the cofermented 'X' component (25), another heptaene closely related to it. The mobile phase consisted of 0.05$\underline{M}$ sodium acetate buffer (pH 5.0)-methanol-acetonitrile(45:35:30) containing 3 m$\underline{M}$ EDTA. Figure 19 shows the separation obtained when a sample of amphotericin B was chromatographed on a 25-cm PRP-1 column under the same chromatographic conditions. The amphotericin B was retained even less (4.6 min.) than that on a shorter C_{18} column (RT 5.5 min.) with marginal separation from the 'X' component. Using a C_{18} column of approximately equal dimensions(25 cm x 4.6 mm i.d.) as the PRP-1 column, a retention time of $\sim$ 20 minutes was obtained for amphotericin B. The poor retention of macrocyclic structures like amphotericin B on PRP-1, compared to bonded reverse-phase, could be explained only by size exclusion. Here, obviously, the selectivity and retention were not as good as that obtained with other modes of separation (adsorption, partitioning, π-π interaction, H-bonding etc.) which are prevalent in poly(styrene-divinylbenzene), PRP-1, stationary phase. This was further exemplified by the poor resolution obtained between amphotericin B and its 'X' component.

Acknowledgments

The author is grateful to Dr. G. Brewer, Dr. B. Kline, and Dr. J. Kirschbaum for their encouragement and many helpful suggestions. The author also wishes to thank Ms. C. Saloom for her conscientious typing of this manuscript and Mr. J. Alcantara for the artwork.

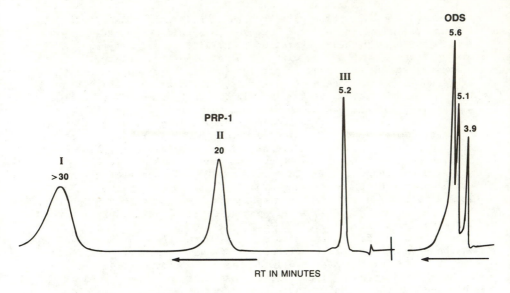

Figure 17.   Comparison of separations on PRP-1 and ODS columns
with mobile phase containing no TBAHS.

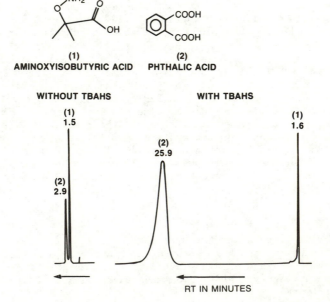

Figure 18.   Effect of TBAHS in the separation of fairly ionic
compounds using a PRP-1 column. Peaks: (1) = aminoxyisobutyric
acid, (2) = phthalic acid.

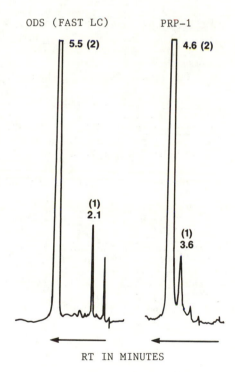

Figure 19.   Retention of amphotericin B on PRP-1 and C$_{18}$ columns. Mobile phase: 45% 0.05M sodium acetate buffer – 3 mM EDTA– 35% acetonitrile – 20% methanol (pH 5.0). Flow rate: 1 mL/min. Detection: 405 nm. Peaks: (1) = amphotericin X, (2) = amphotericin B.

Literature Cited

1. Snyder, L.R.; Kirkland, J. J. "Introduction to Modern Liquid
    Chromatography"; 2nd Edition. John Wiley and Sons: New York,
    1979; Chap. 7-9.
2. Majors, R.E. J. Chromatogr. Sci. 1980, 18,488-11.
3. Kroeff, E.P.; Pietrzyk, D.J. Anal. Chem. 1978, 50, 502-11.
4. Roueliotis, P.; Unger, K.K. J. Chromatogr. 1978, 149,211-24.
5. Cooke, N.H.C.; Olsen, K. J. Chromatogr. Sci. 1980, 18, 512-24.
6. Benson, J.R.; Woo, D.J. J. Chromatogr. Sci. 1984, 22, 386-99.
7. Engelhardt, H.; Muller, H. J. Chromatogr. 1981, 218, 395-07.
8. Lee, D.P. J. Chromatogr. Sci. 1982, 20, 203-08.
9. Alton, K.B.; Lietz, F.; Bariletto, S.; Jaworsky, L.;
    Desrivieres, D.; Patrick, J. J. Chromatogr. 1984, 311, 319-28.
10. Lee, D.P.; Kindsvater, J.H. Anal. Chem. 1980, 52, 2425-28.
11. Greyson, R.; Patch , A. M. J. Chromatogr. 1982, 242, 349-52.
12. Buta, J. G. J. Chromatogr. 1984, 295, 506-09.
13. Gupta, R.N.; Smith, P.T.; Eng, F. Clin. Chem. 1982, 28,
    1772-74.
14. Bontemps, J.; Bettendorft, L.; Lombet, J.; Grandfils, C.;
    Dandrifosse, G.; Schoeffeniels, E. J. Chromatogr. 1984, 295,
    486-91.
15. Lee, D.P. J. Chromatogr. Sci. 1984, 22, 327-31.
16. Sykes, R.B.; Bonner, D.P.; Bush, K.; Georgoadakou, N.H.
    Antimicrob. Agents Chemother. 1982, 21, 82-92.
17. Cimarusti, C.M.; Applegate, H.E.; Chang, H.W.; Floyd, D.M.;
    Koster, W.H.; Slusarchyk, W.A.; Young, M.G. J. Org. Chem.
    1982, 47, 180-82.
18. Joseph, J.M.; Kirschbaum, J.J. Abstr. 131st Apha Annual
    Meeting. 1984, P.92.
19. Cantwell, F.F.; Puon, Su. Anal. Chem. 1979, 51, 623-32.
20. Kroeff, E.P.; Pietrzyk, D.J. Anal. Chem. 1978, 50, 502-11.
21. Wood, R.; Jupille, T. J. Chromatogr. Sci. 1980, 18, 551-58.
22. Bidlingmeyer, B.A. J. Chromatogr. Sci. 1980, 18, 525-39.
23. Horvath, C.; Melander, W.; Molnar, I. Anal. Chem. 1977, 49,
    142-54.
24. Rotsch, T.D.; Cahill, Jr., W.R.; Pietrzyk, D.J.; Cantwell, F.F.
    Can. J. Chem. 1981, 59, 2179-83.
25. Margosis, M.; Aszalos, A. J. Pharm. Sci. 1984, 73, 835-38.

RECEIVED April 19, 1985

Chromatographic Separation of Enantiomers on Rationally Designed Chiral Stationary Phases

William H. Pirkle

School of Chemical Sciences, University of Illinois, Urbana, IL 61801

The separation of enantiomers by liquid chromatography is now feasible on both analytical and preparative scales. Recent developments in chiral stationary phase (CSP) design have extended the scope of enantiomer separations to a point unimagined a few years ago. The enantiomers of literererally tens of thousands of compounds can now be separated chromatographically, often with considerable understanding of the mechanism of the separation process. Such understanding enhances one's ability to design improved CSPs and to assign absolute configurations from observed elution orders.

It has long been understood that chromatography of racemates on chiral adsorbents might result in enantiomer separation. Initial efforts to use convenient chiral adsorbents (cellulose, starch, wool) usually met with scant success. With the advent of more sophisticated techniques, it has become possible to separate the enantiomers of a large number, literally tens of thousands, of compounds by liquid chromatography. Because of the close ties between life and chirality, the ability to separate enantiomers by HPLC has an eager audience of potential users.

Since enantiomer separation requires the intervention of some chiral agent, one may utilize either chiral mobile phase additives (CMPA) or chiral stationary phases (CSPs). While the requirement that one add a chiral substance to the mobile phase has obvious limitations for preparative separations, it is not a serious problem for analytical separations. Indeed, for some types of compounds (e.g. amino acids) this approach may be preferred. Quite an extensive literature exists for the use of mobile phases containing chiral bidentate ligands and copper ions for the "ligand exchange" resolution of underivatized amino acids (1,2) and for N-dansyl derivatives of amino acids (3,4). Tartaric acid derivatives have also been used as CMPAs (5).

When one uses CMPAs, there is always as question as to why enantiomer separation occurs (if it does). The same is true for those CSAs that we may term "biopolymers". This constitutes a

0097–6156/86/0297–0101$06.00/0
© 1986 American Chemical Society

problem of sorts although one needn't understand why a separation occurs in order to use the separation. Once enantiomers are caused to give rise to separate chromatographic peaks, area measurement affords the enantiomeric purity of the sample. However, if one doesn't understand why a separation should occur, then one is reduced to a "trial and error" search for a system capable of producing the desired result.

A great many interesting and useful enantiomer separations have been performed on "biopolymer" CSPs, several of which will be mentioned for they are commercially available. Columns containing swollen triacetylated microcrystalline cellulose, available from Merck, have been extensively investigated by Mannschrek (6,7) and used in both analytical and preparative modes. Although there is little detailed understanding as to how and why this CSP works, it is thought that the laminar nature of the swollen crystals offers chiral cavities into which enantiomers must intercalate. The recent cyclodextrin CSPs, available from Advanced Separations Technology, clearly require intercalation into the hole in the chiral cyclodextrin (8). Beyond this, understanding is dim.

Still less well defined is the mode of action of various cellulose derivatives coated onto diphenyl-silanized silica (9,10). Daicel markets columns made from actylated, benzoylated, cinnamoylated, phenyl carbamoylated and benzylated celluloses. While these columns afford a number of interesting enantiomer separations, there is no clear pattern as to what will resolve on which column nor in what order the enantiomers will elute. Consequently, the design of these CSPs is approached empirically and their use is on a trial and error basis. A variety of polysaccharides has been so investigated. Again, no clear pattern of performance is evident (10).

Columns packed with silica-bound proteins have been devised; bovine serum albumin-derived columns (11) are available from Macherey Nagel. These CSPs show a fairly extensive scope of action, although the nature of the chiral recognition processes employed is still vague. Owing to the low concentration of active sites, these columns require quite small samples (0.5 - 5 nmol). Mobile phase variation is possible (over a limited range) and alters the chromatographic behavior of the enantiomers markedly. The columns must be treated carefully so as to not destroy the protein CSP.

The biopolymer CSPs are attractive owing to the ready availability of the chiral precursors. Offsetting this advantage is an innate complexity that more or less baffles one's ability to deduce the details of the operative chiral recognition processes. Moreover, one cannot easily alter or "fine tune" the structure of the CSP to enhance selectively. Finally, there may well be limitations as to the mobile phases which can be used owing to possible swelling, shrinking, denaturation, or dissolution.

A great deal of the complexity of polymeric CSPs stems from the analyte perceiving the CSP as a chiral array of subunits (monomers) which may themselves be chiral. Unless one knows the structure of the array, one will have difficulty in specifying a possible chiral recognition mechanism. Even then, an abundance of closely spaced potential interaction sites can preclude mechanistic understanding. This is also true for the synthetic

polymeric CSPs which have been devised. The more noteworthy of
these is that of Blaske, derived from the acrylamide of
α-phenylethylamine (12) and the chiral polymer of Okamoto, derived
from the triphenylmethyl ester of methacrylic acid (13). The
latter is available from Daicel. Both phases perform useful
separations and are thought to work by inclusion-like processes.

An alternative approach (and one we have favored) is to
consider chiral recognition as it might occur between small chiral
molecules in solution. Chiral recognition cannot occur unless
there are at least three simultaneous interactions, at least one
of which is stereochemically dependent, between a chiral
"recognizer" and one of the enantiomers whose configuration is to
be "recognized." The interactions so employed are the usual
intermolecular interactions which occur in solution. One simply
engineers a chiral molecule that contains functional groups
capable of undergoing these interactions, anchors it to a support,
packs a column, and chromatographs a racemate which contains
appropriate functionality. Complimentary functionality must be
present if the analyte is to undergo the required multiple
simultaneous interactions. This simple rationale has great value
and far-reaching implications. It provides an a priori basis for
CSP design and it provides a means of understanding the
subsequently observed chromatographic behavior of assorted
racemates on that CSP. It provides a means for rationally
improving the performance of a given CSP. Finally and
importantly, the CSP is synthetic and its structure can be altered
and controlled at will.

For example, our first CSP was 9-anthryl trifluoromethyl
carbinol, linked at the 10-position to silica through a six atom
connecting arm. This CSP was intended to utilize π-π
interactions, hydrogen bonding interactions, and steric
interactions to separate the enantiomers of compounds containing
π-acidic and basic sites (14). Among the analytes which are part
of this clientele are N-(3,5-dinitrobenzoyl) derivatives of amines
and amino acids. Chiral recognition is reciprocal in that if a
CSP derived from (+)-A selectively retains (+)-B, then a CSP
derived from (+)-B should selectively retain (+)-A. On this
basis, CSPs derived from N-(3,5-dinitrobenzoyl) amino acids were
prepared and evaluated (14,15).

The N-(3,5-dinitrobenzoyl) amino acid-derived CSPs show
extraordinary scope and follow readily understandable patterns of
behavior. These CSPs have been commercialized; analytical and
preparative columns and packings are available from the Regis,
J. T. Baker, and Sumitomo companies. The structural requirements
these CSPs exact from resolvable analytes are stipulatable, albeit
still in imprecise terms. For example, the N-(3,5-dinitrobenzoyl)
amino acid columns utilize combinations of π-π; hydrogen bonding,
dipole-dipole, and steric interactions to achieve chiral
recognition. Thus, the analyte must contain a combination of
π-basic, basic or acidic "sites," possibly have a strong dipole,
or perhaps contain a bulky steric interaction site. These
interaction sites must be arranged so as to act in concert.
Generalized (16,17) and specific chiral recognition models have
been presented (18) for a variety of analyte types.

Large columns of the N-(3,5-dinitrobenzoyl) amino acid CSPs have been prepared and multigram quantities of racemates have been separated automatically (19). Fifty grams of racemic methyl N-(2-naphthyl) alaninate have been resolved per pass on a 6" X 4' column containing 13 kg of chiral packing.

Among the analyte enantiomers separable on the N-3,5-dinitro-benzoylphenylglycine CSPs are N-acylated α-arylalkylamines (20,21). After a study of the relation between analyte structure, chromatographic separability, and operative chiral recognition mechanism(s), several reciprocal α-arylalkylamine-derived CSPs were designed and constructed (22). These CSPs afford excellent selectivity for the N-3,5-dinitrobenzoyl derivatives of amines, α-amino esters, amino alcohols and α-aminophosphonates. From systematic study of such analytes, a deeper understanding of chiral recognition processes has been attained and employed to further improve the design of the α-arylalkylamine-derived CSPs. As a consequence, unusually large separability factors are noted; values ranging between 2 and 8 are considered typical (16,17). Large separability factors facilitate preparative separations. These CSPs are not yet commercially available. Somewhat similar but less optimized (hence, less selective) CSPs are available from Sumitomo. These CSPs stem from the work of Ōi et al and have many practical applications (23). They are derived from N-acylated 1-(α-naphthyl)ethylamine, commercially available from several sources.

Similar CSPs derived from linking α-arylalkylamines to silica through urea functionality have been prepared by Ōi (23,25) and by ourselves (26). These CSPs are workable but usually inferior to the corresponding amide-linked CSPs. A commercial urea-linked α-phenylethylamine CSP column is available from Supelco but will almost always show considerably less selectivity and narrower scope than the more sophisticated amide-linked CSPs derived from more "optimized" α-arylalkylamines.

One important aspect of the α-arylalkylamine CSPs, both amide and urea-linked, is that they often have more than one chiral recognition process available to them. It was recently shown that these CSPs utilize two competing processes of opposite enantioselectivities (16,17). Optimization entails not only altering the structure of the chiral entity so as to maximize the strengths of essential interactions, it also entails the manner in which the chiral entity is connected to the silica support and the spacing between adjacent strands of bonded phase (27). In simple terms, one process is more intercalative than the other. Densely packed strands disfavor the intercalative process, thereby enhancing the contribution of the nonintercalative process. Similarly, the orientation of the chiral entity with respect to the silica surface determines whether, using a given combination of analyte-CSP interactions, a portion of the analyte is intercalated between adjacent strands. In organic mobile phases, intercalation can lead to steric repulsion. Thus, different orientations of the chiral entity with respect to the silica will alter the relative contributions of the two competing processes. By largely suppressing one process, rather high selectivity may be obtained from the other. In aqueous mobile phases, lipophilic interaction begins to compensate for the steric repulsion which

attends intercalation. Although mechanistically rather
interesting, the practical consequence of use of aqueous mobile
phases is usually a reduction in selectivity. However,
selectivity is still great enough for high quality analysis of
enantiomeric purity.

In the area of enantiomer separation, the number of available
CSPs is rapidly proliferating. While this may be confusing, it
does offer the researcher an opportunity to perform a number of
"trial and error" experiments to effect the desired resolution.
Among these proliferating CSPs are some whose mode of interaction
is relatively well understood. This offers the researcher a way
to rationally match a CSP to his analyte so as to maximize the
probability of success.

Literature Cited

1. Davankov, V. A.; Kurganov, A. A.; Bochkov, A. S.; in
    "Advances in Chromatography, Vol. 22"; Giddings, J. C.;
    Grushka, E; Cazes, J. and Brown, P. R. Ed.; Marcel Dekker,
    New York, 1984; p 71-116.
2. Weinstein, S.; Engel, M. H.; Hare, P. E. Anal.Biochem. 1982,
    121, 370-377.
3. Weinstein, S. Tetrahedron Lett. 1984, 985-9.
4. Feibush, B.; Cohen, M. J.; Karger, B. L. J. Chromatogr.
    1983, 282, 3-26
5. Dobashi, A.; Hara, S. Anal Chem. 1983, 55, 1805-6.
6. Koller, H.; Rimbock, K. H.; Mannschreck, A. J. Chromatogr.
    1983, 282, 89-94.
7. Mannschreck, A; Koller, H.; Wernicke, R. Merck Kontakte H1
    1984,   2-14.
8. Hinze, W. L.; Riehl, T. E.; Armstrong, D. A.; DeMond, W.;
    Alak, A.; Ward, T. Anal. Chem. 1985, 57, 237-242.
9. Okamoto, Y; Kawashima, M; Hatada, K. Chem Lett. 1984, 739.
10. Okamoto, Y.; Kawashima; Hatada, K. J. Am. Chem. Soc. 1984,
    106, 5357-9.
11. Allenmark, S. LC 1985, 3, 348-353.
12. Schwanghart, A.; Backmann, W.; Blaschke, G. Chem. Ber. 1977,
    110, 778.
13. Yuki, H; Okamoto, Y.; Okamoto, I. J. Am. Chem. Soc. 1980,
    102, 6356.
14. Pirkle, W. H.; House, D. W.; Finn, J. M. J. Chromatogr.
    1980, 192, 143-158.
15. Pirkle, W. H.; Finn, J. M.; Schreiner, J. L.; Hamper, B. C.
    J. Am. Chem. Soc. 1981, 103, 3964-6.
16. Pirkle, W. H.; Hyun, M. H.; Bank, B. J. Chromatogr. 1984,
    316, 585-604.
17. Pirkle, W. H.; Hyun, M. H.; Tsipouras, A.; Hamper, B. C.;
    Bank, B. J. Pharm.Biomed. Anal. 1984, 2, 173-181.
18. Pirkle, W. H.; Finn, J. M.; Hamper, B. C.; Schreiner, J.;
    Pribish, J. R. in "Asymmetric Reactions and Processes in
    Chemistry, No 185"; Eliel, E. and Otsuka, S. Ed.; American
    Chemical Society: Washington, D.C., 1982.
19. Pirkle, W. H.; Finn, J. M. J. Org. Chem. 1982, 47,
    4037-4040.
20. Pirkle, W. H.; Welch, C. J. J. Org. Chem. 1984, 49, 138-140.

21.   Pirkle, W. H.; Welch, C. J.; Hyun, M. H.   J. Org. Chem. 1983,
      48, 5022-5026.
22.   Pirkle, W. H.; Hyun, M. H.   J. Org. Chem. 1984, 49,
      3043-3046.
23.   Ōi, N.; Nagase, M.; Doi, T.   J. Chromatogr. 1983, 257,
      111-117.
24.   Ōi, N.; Kitahara, H.; Doi, T; Yamamoto, S.   Bunseki Kagaka
      1983, 32, 345.
25.   Ōi, N.; Kitahara, H.; "Abstracts of the 49th Biannual Meeting
      of the Chemical Society of Japan, I", 1984, 386.
26.   Pirkle, W. H.; Hyun, M. H.   J. Chromatogr. 1985, 322,
      295-307.
27.   Pirkle, W. H.; Hyun, M. H.   J. Chromatogr. 1985, in press.

RECEIVED May 6, 1985

Nonoptical Noise Sources in High-Performance Liquid Chromatographic Optical Absorbance Detectors

Seth R. Abbott and Herman H. Kelderman

Varian Instrument Group, Walnut Creek Division, Walnut Creek, PA 94598

Improvement of optical detection limits can be achieved by either increasing the signal or decreasing the noise. In an ideal optical detector, noise is determined by the fundamental "shot" or statistical noise of the photon flux incident on the photodetector.

However, this ideal situation is not always realized in HPLC optical detectors, for which both electronic and thermal noise sources can exceed optical shot noise, degrading the signal to noise level inherent to the optical design of the detector.

Such "non-optical" noise sources are significant in current HPLC optical absorbance detectors in that they limit the reduction of fixed wavelength detector noise below approximately 10^{-5} absorbance units (au) and limit the noise achievable in deuterium lamp-based photodiode array detectors to approximately 4×10^{-5} au. (The above noise values are given as peak to peak noise as seen on a recorder, which is approximately 6 times the rms noise for the case of a Gaussian noise distribution (1).

Reduction of "non-optical" noise sources below the shot noise limit of the detector is thus critical in extending the performance of HPLC optical detectors. One must understand these noise sources prior to optimizing the design of a detector, and prior to applying new technology such as that of the linear photodiode array to HPLC detection.

Noise Categories

The noise of an optical detector is the root mean square of a series of individual optical, electronic and thermal noise source contributions (assuming that the noise sources are uncorrelated).

$$N = \sqrt{\Sigma N_i^2} = \sqrt{\Sigma N_{opt}^2 + \Sigma N_{elec}^2 + \Sigma N_{thermal}^2} \qquad (1)$$

Consider the case of an optical absorbance detector. Each individual noise current n_i translates into an absorbance noise N_i through the logarithmic relationship:

0097–6156/86/0297–0107$06.00/0
© 1986 American Chemical Society

$$N_i \; (au) = \log \left[1 + \frac{n_i}{I_s} \right] \qquad I_s = \text{photodetector current} \qquad (2)$$

where the term n_i/I_s is denoted as the noise fraction F_i. One can see that for a noise source to contribute $\lesssim 10^{-5}$ au to a detector noise N, the noise fraction must be held to approximately 20 ppm. This places stringent requirements on optical source stability and referencing schemes, and on the electronic and thermal stability of the detector components and electromechanical design.

In considering a detector design, one can estimate the photodetector current I_s through a knowledge of the source intensity and the collection efficiency of the optical system (including the critical flow cell). One can then calculate N_i terms for various optical, electronic and thermal sources using standard equations or experimental measurements. These are then translated into N_i terms, from which one determines the relative contribution of each noise source to a system noise N. The result of this calculation then guides the detector design effort as to which noise source must be reduced to attain optimum performance.

From the viewpoint of detector design, it is instructive to classify noise sources in terms of the dependence of the N_i term on the photodetector current I_s. (Note that the absorbance signal A of an optical absorbance detector is not dependent on I_s.) Common noise sources can thus be divided into three categories:

Category I. The noise current n_i is dependent on I_s, and the absorbance noise contribution N_i is dependent on I_s:

$$N_i = \log \left(1 + \frac{f \; (I_s)}{I_s} \right) = g \; (I_s) \qquad (3)$$

Optical and electronic shot noise belong to this category, with $n_i \alpha \sqrt{I_s}$ and hence $N_i \; \alpha 1/\sqrt{I_s}$.

Category II. The noise current n_i is independent of I_s, and the absorbance noise contribution N_i is dependent on the first power of I_s:

$$N_i = \log \left(1 + \frac{n_i}{I_s} \right) = f(I_s) \text{ where } n_i \neq f(I_s) \qquad (4)$$

Diode array readout (scanning) noise and other electronic noises such as Johnson thermal noise in a readout resistor and photodetector dark current noise belong to this category.

Category III. Noise source n_i is a fixed fraction of I_s, such that the absorbance noise contribution N_i is independent of the magnitude of I_s.

$$N_i = \log \left(1 + \frac{k \; I_s}{I_s} \right) = \log \; (1 + k) \neq f(I_s) \qquad (5)$$

These noise sources are due to changes in flow cell optical throughput and changes in photodetector response caused by thermal fluctuations. Absorbance noise due to pump pulsations belongs to this category, as does noise due to the difference in the thermal coefficients and thermal environment of sample and reference photodetectors. The former "pulse" noise source is a critical limit in

HPLC analyzers based on post-column reaction detection. However PCR noise sources will not be addressed further in this report.

The three major noise categories are summarized in Table I, below:

Table I.   Optical Detector Noise Categories.

CATEGORY	CHARACTERISTICS	TYPES
I.	$n_i = f(I_s)$ $N_i = g(I_s)$	1. Optical shot noise 2. Electronic shot noise
II.	$n_i \neq f(I_s)$ $N_i = f(I_s)$	3. Johnson thermal noise in resistor 4. Photodetector dark current noise 5. Array readout (scanning) noise
III.	$n_i = f(I_s)$ $N_i \neq f(I_s)$	6. Thermal noise due to sample and reference diode response temperature coefficients. 7. Thermal noise due to mobile phase/flow cell temperature coefficient. 8. Pump pulsation noise.

Category I noise is reduced by a design strategy focussing on increasing the photodetector current I_s through either increased source intensity and/or increased optical collection efficiency. An x-fold increase in I_s will yield a $\sqrt{x}$ reduction in absorbance noise. Category II noise is reduced by a combination of strategies in which n_i is reduced through appropriate electronic and thermo-electronic design and through increasing the photodetector current I_s. Note that in this case, an x-fold increase in I_s yields a full x-fold reduction in absorbance noise.

Category III noise is reduced by strategies combining optimization of both detector and non-detector hardware as in reduction of reciprocating pump pulsation, the use of low thermal coefficient, matched pairs of photodiodes for reference and sample beam optical trains, and thermal coupling of reference and sample photodiodes. Pump pulsation noise can also be reduced by proper flow cell design (2).

Noise Characteristics and Equations

Category I Noise Sources.

1. Optical Shot Noise.   The optical shot noise current of a given photon flux is proportional to the square root of the number of photons detected in a given time period:

$$\text{rms } n_{os} = [2q\ I_s\ B]^{1/2} \tag{6}$$

where τ is the coulometric charge of an electron and the noise "bandwidth" B is equivalent to $1/2\ \pi\tau$, where τ is the time constant of the detector amplifier circuit. Combining Equations 2 and 6, one obtains

$$\text{rms } N_{os} = \log \left[1 + \left(\frac{2qB}{I_s} \right) \right]^{1/2} \tag{7}$$

For HPLC optical absorbance detectors, the term $(2qB/I_s)^{1/2}$ $\ll 1.0$ and thus an x-fold increase in I_s will reduce the absorbance noise contribution N_{os} by $\sqrt{x}$.

2. Electronic Shot Noise. The electronic shot noise of an electron current I_s flowing through a readout or feedback resistor of a detector amplifier circuit is proportional to the square root of the number of electrons passing through the resistor in a given time period:

$$\text{rms } n_{es} = [2q\, I_s\, B]^{1/2} \tag{8}$$

Thus in detecting a given photon flux and converting it into a signal current I_s, one incurs an inherent shot noise current n_s such that:

$$\text{rms } n_s = (n_{os}^2 + n_{es}^2)^{1/2} = n_{os} \sqrt{2} \tag{9}$$

For a state of art variable wavelength UV absorbance detector based on a deuterium lamp source and a 5 nm spectral bandwidth, I_s is of the order of 25 nanoamps. For $T=0.5$ seconds, Equations 2, 6, 8, 9 evaluate to an rms shot noise current n_s of 7×10^{-14} amps, an rms shot noise fraction F_s of 2.9×10^{-6} and a peak to peak absorbance shot noise contribution N_s of 7.8×10^{-6} au.

This value is approximately 2-fold less than the typical noise of 1.5×10^{-5} au observed for a state of art variable wavelength detector,* suggesting that a significant non-optical noise source or sources are present.

For a state of art photodiode array UV absorbance detector based on a deuterium lamp source and 5 nm spectral bandwidth, I_s is of the order of ≤ 3 nanoamps. For $T=0.5$ seconds, Equations 2, 6, 8, 9 evaluate to an rms shot noise current n_s of 1.8×10^{-14} amps, an rms noise fraction F_s of 5.8×10^{-6} and a peak-to-peak absorbance noise contribution N_s of 2.2×10^{-5} au. This is approximately 2-8 fold less than the noise of current array detectors, indicating the presence of a major non-optical noise source or sources. This is an electronic noise source, called array readout noise (4), which will be described in a later section.

---

*Note that shot noise in this treatment was considered for the case of the sample diode channel only. R ference schemes in which a beam splitter is placed before the dispersive element (filter or mono-chromator) yield reference channel currents significantly larger than I_s, so that the contribution of the reference shot noise is minimal.

Category II Noise Sources.

This category encompasses electronic and thermoelectronic noise sources of great importance to the application of silicon photodiodes and photodiode arrays to optical detection in HPLC.

3. Johnson thermal noise. Noise generated by thermal agitation of electrons (electron kinetic energy fluctuations) in a resistor sets a lower limit on noise present in a detector readout circuit. Called Johnson thermal noise, this non-periodic noise exists in all conductive elements, and is a function of temperature and resistance, described by:

$$\text{rms } n_J' = \left(\frac{4\ kT\ B}{R}\right)^{1/2} \tag{10}$$

where k is the Boltzmann constant. Since Johnson thermal noise arises in both sample and reference channels of an optical detector,

$$\text{rms } n_J = n_J' \sqrt{2} \tag{11}$$

At typical ambient temperature (300°K), and 0.5 sec τ, in a typical megohm load resistor across each photodiode, Equations 10 and 11 evaluate to rms n_J value of 9.9×10^{-14} amps, such that for the variable wavelength detector with an I_s of 25×10^{-9} amps, the Johnson noise fraction evaluates to an rms noise fraction F_J of 4×10^{-6}, generating a Johnson absorbance noise contribution N_J (peak to peak) of 1.0×10^{-5} au. This value is slightly in excess of the variable wavelength detector shot noise (N_s) calculated above and thus should be considered in state of art optical absorbance detector design.

4. Photodetector dark current noise. The "dark current" of a photodetector refers to current which flows through that photodetector at zero incident light level, and is due to a variety of thermoelectronic sources. This produces a dark current noise in sample and reference channels described by

$$\text{rms } n_{dc}' = (2q\ i_{dc}\ B)^{1/2} \text{ and} \tag{12}$$

$$n_{dc} = n_{dc}' \sqrt{2} \tag{13}$$

where the dark current i_{dc} is a function of temperature, approximately doubling with every 10°C rise in temperature (3).

For a UV-sensitive silicon photodiode near ambient temperature, $i_{dc} \leq 2 \times 10^{-11}$ amps, and for $\tau = 0.5$ sec, eq. (13) evaluates to an rms n_{dc} of 2.0×10^{-15} amp.

Again, for the variable wavelength detector with I_s of 25×10^{-9} amps, one calculates a diode dark current rms noise fraction F_{dc} of 8×10^{-8} and a dark current absorbance noise contribution N_{dc} (peak to peak) of 2.1×10^{-7} au. This is not a significant limit to current absorbance detectors. However it is a significant limit to the application of photodiodes to fluorescence detection, for which I_s is 1000-fold lower than in absorbance detection.

5. **Photodiode array readout (scanning) noise.** In a photodiode array, the photon flux incident on each diode causes charge to develop across the diode capacitance C_d (seen in the schematic of Figure 1). At an integration time interval determined by the scan rate of the array, a JFET readout switch is closed, dumping the photoelectrically generated charge of C_d onto a readout capacitor C_1. The charge is then read out as a voltage at the output of an operational amplifier.

The JFET switch is closed by the step application of a −5 volt signal. Whenever this voltage step is applied to the open switch, the inherent junction capacitance of the JFET causes charge to be injected into the readout circuit. The fluctuation in this charge generates an electronic readout or scanning noise current (n_a), which is independent of the photodiode signal current I_s.

Array readout noise is minimized by reducing the switch junction capacitance to approximately 1-2 picofarads (this value can be higher on certain commercial self-scanned arrays) and by closely regulating the −5 volt power supply. Noise measurements on experimental array detectors indicate a readout noise N_a (peak to peak) of approximately 4×10^{-5} au for $T = 0.5$ sec. Since reverse optic polychromator/array designs yield I_s values of $\lesssim 3$ nanoamps, one can estimate an rms n_a of approximately 1.6×10^{-5} amps. This noise significantly limits the performance of current array detectors.

The magnitude of the array readout noise will also severely limit the application of the photodiode array to HPLC fluorescence detection since the fluorescence detector background current I_s is typically 1000-fold less than that of an absorbance photo-detector current. The low I_s value in fluorescence magnifies the relative effect of Category II noise sources such as array readout noise.

Category III   Noise Sources.

6. **Noise due to fluctuations in thermal environment and the temperature coefficient of photodiode response.** The quantum efficiency of a photodiode is a function of temperature. A state of art UV-sensitive photodiode with a typical quantum efficiency of 75% has a temperature coefficient of between 0.02 to 0.2% per $^\circ$C near ambient temperature (5). Use of sample and reference diode pairs that are experimentally matched in terms of response temperature coefficient, can reduce this noise to approximately 4×10^{-5} au (peak to peak) per $^\circ$C. Short term temperature fluctuations within optical detectors are of the order of 0.1°C and thus this "tempco" noise N_t is approximately 4×10^{-6} au.

7. **Mobile phase/flow cell temperature coefficient noise.** Short term fluctuations in flow cell temperature can result in absorbance noise due to (1) a change in the flow cell refractive index profile (thermal lens) which can alter flow cell light throughput; and (2) an actual change in the absorption spectrum of the mobile phase, an effect seen only if one operates near or below the UV cutoff wavelength of the mobile phase. The former effect is minimized by an optical design in which incident light clears the flow cell **without** striking the flow cell wall/liquid interface region and by heat sinking the mobile phase inlet line to the flow cell body. Experimental measurements using such an optical design

and $\lambda \gtrsim 205$ nm, indicate a tempco noise N_m of $<5 \times 10^{-5}$ au/°C. Flow cell short term thermal fluctuations are typically $\lesssim 0.1$°C and thus the mobile phase/flow cell tempco noise N_m is $<5 \times 10^{-6}$ au.

Pump Pulsation Induced Noise. So-called "pulse" noise has a period matching that of the reciprocating pump fill stroke, and is caused by a change in the flow cell's thermal lens. A Fourier transform power spectrum showing absorbance detector pulse noise is shown in Figure 2. This effect is treated in detail elsewhere.[2] Proper flow cell design, identical to that required to reduce mobile phase tempco noise, yields an absorbance detector flow sensitivity of $\lesssim 2 \times 10^{-4}$ au per mℓ/min flow change for organic mobile phase and approximately 3-fold lower for aqueous mobile phase. State of art HPLC reciprocating pump-based solvent delivery systems reduce pump pulsations to $\leq 2\%$ and hence one obtains a peak to peak noise N_p of $<4 \times 10^{-6}$ au. Since reciprocating pumps utilize fast fill periods, the detector time constant can act to reduce pulse noise, such that with a $\tau = 0.5$ sec, pulse noise is $\lesssim 2 \times 10^{-6}$ au (peak to peak).

Variable Wavelength UV Absorbance Detector Noise Summary

It is now possible to consider the contribution of the previously described noise sources to the system noise N of optical absorbance detectors. Consider a state of art variable wavelength UV absorbance detector, based on a 30 watt deuterium source, a 5 nm spectral bandwidth, and a UV-sensitive silicon diode photodetector. At 250 nm, the deuterium source geometry allows 113 watts through an étendue of 0.63 steradian-mm^2. A 5$\mu\ell$ flow cell with a 4 mm path and an optical train designed such that incident light clears the flow cell walls allows collection of 0.094 steradian-mm^2, which is accommodated by a small Ebert monochromator. The efficiency of the entire optical train is 0.06 and thus the sample photodiode should collect $(0.094/0.63) \times 0.06 \times 113$ μwatts = <u>1.0 μwatt</u>, corresponding to a 100 nanoamp sample current (I_s). In practice, 25 nanoamp is achieved with this design in its manufactured (real) state.

Assuming a detector τ of 0.5 sec, and using the treatment of Section III, one estimates the noise source contributions listed in Table II, below:

Table II.   Noise Source Contributions – Variable
Wavelength Detector ($\tau = 0.5$ sec)

1, 2.	Shot noise	N_s	7.8×10^{-6} au
3.	Johnson thermal noise ($10^6 \Omega \, R_f$)	N_J	1.0×10^{-5} au
4.	Photodiode dark current noise	N_{dc}	2.1×10^{-7} au
6.	Sample/reference diode Tempco noise	N_t	4.0×10^{-6} au
7.	Mobile phase/flow cell Tempco noise	N_m	$<5 \times 10^{-6}$ au ($\lambda < 205$ nm)
8.	Pump pulse noise	N_p	$\lesssim 2 \times 10^{-6}$ au

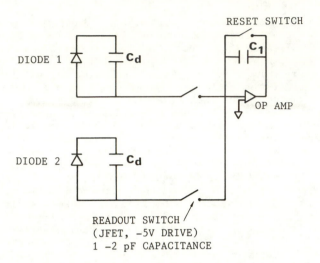

Figure 1. Schematic diagram of photodiode array readout circuit.
C_d = diode capacitance. C_1 = readout capacitor.

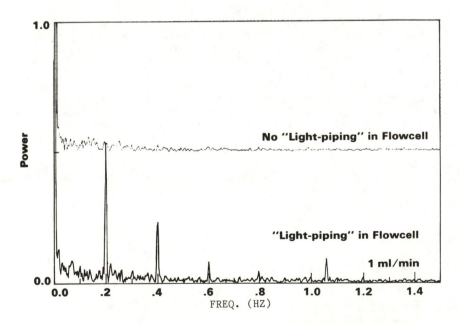

Figure 2. Fourier transform of UV absorbance detector baseline
for mobile phase flow rate of 1 mℓ/min water, corresponding to
0.2 Hz reciprocating piston frequency. The upper trace is the
transform for the detector in which incident light clears the
flow cell walls. The lower trace is the transform for a detector
in which incident light strikes the flow cell walls ("light
piping").

Thus, shot and Johnson thermal noise are the major noise determinants of this system. Application of Equation 1 predicts a system noise N of 1.43×10^{-5} au, (peak to peak) close to the experimentally measured noise (1.5×10^{-5} au) of such a detector.

Fixed Wavelength Detector Noise Summary

Now, one must consider how to improve optical absorbance detection limits, that is - how to reduce noise. One strategy is to increase I_s through use of a fixed wavelength detector based on a discrete line lamp and an isolation filter to select the UV line of interest generated by the lamp (typically Zn, Cd or Hg lamps having major lines at 214 nm, 229 nm, 254 nm, respectively).

For the case of a 5 watt Hg lamp, at 254 ± 0.01 nm, the source geometry allows 476 watts through an étendue of 0.63 steradian-mm^2. A $5\mu\ell$ flow cell with a 2.5 mm path and an optical train designed such that incident light clears the flow cell walls should allow collection of the full 0.63 steradian-mm^2, which is accommodated by an isolation filter. The efficiency of the entire optical train is 0.12 and thus the sample photodiode should collect $0.63/0.63 \times 0.12 \times 476$ μwatt$=57$ μwatt. Since the variable wavelength detector treatment of Section IV overestimated the actual I_s by 4-fold, assume that the fixed wavelength detector allows an I_s corresponding to $57/4=14.3$ μwatt or 1.43 microamps.

Assuming a detector τ of 0.5 sec, and using the treatment of Section III, one estimates the noise source contributions listed in Table III below:

Table III. Noise Source Contributions - Fixed Wavelength Detector (254 nm, $\tau=0.5$ sec).

1, 2.	Shot noise	N_s	1.0×10^{-6} au
3.	Johnson thermal noise ($10^6 \Omega$ R_f)	N_J	1.8×10^{-7} au
4.	Photodiode dark current noise	N_{dc}	3.7×10^{-9} au
6.	Sample/reference diode Tempco noise	N_t	4.0×10^{-6} au
7.	Mobile phase/flow cell	N_m	$<5 \times 10^{-6}$ au
8.	Pump pulse noise	N_p	$\leq 2 \times 10^{-6}$ au

Application of Equation (1) predicts a system noise N of 7×10^{-6} au. Although the particular optical design described has not been reduced to practice, the theoretical estimate of 7×10^{-6} au peak to peak noise is within a factor of 2 of the typical noise ($1-2 \times 10^{-5}$ au) observed on commercial state of art fixed wavelength detectors at 254 nm.

In considering the noise sources of the fixed wavelength detector, note that despite an 8-fold reduction in shot noise relative to the variable wavelength detector of Section IV, only a 2-fold noise reduction is predicted in Table III. What has happened? The increased optical throughput and hence increased value of I_s has significantly reduced both shot noise (Category I) and Johnson thermal and dark current noise (Category II), such that these noise sources are $<10^{-6}$ au. However, the Category III noise source contributions are independent of I_s and have become major noise determinants.

It thus becomes clear that to achieve the shot noise limit (10^{-6} au) of such a fixed wavelength detector, one must reduce the Category III noise sources, posing a challenge to both thermomechanical detector design and pump design.

Photodiode Array UV Absorbance Detector Noise Summary

Consider a UV absorbance detector based on a silicon photodiode array, a reverse optics/polychromator design utilizing a $5\,\mu\ell$ flow cell and a 30 watt deuterium lamp, and a time constant set at 0.5 sec. An experimental model yields a I_s value of $\lesssim 3$ nanoamps. Using the treatment of Section III, one estimates the noise source contributions listed in Table IV below:

Table IV.   Noise Source Contributions – Photodiode
             Array Absorbance Detector

1, 2.	Shot noise	N_s	2.2×10^{-5} au
4.	Photodiode dark current noise	N_{dc}	2.9×10^{-7} au
5.	Array readout	N_a	4.0×10^{-5} au
6.	Sample/reference diode Tempco/noise	N_t	4.0×10^{-6} au
7.	Mobile phase/flow cell noise	N_m	$<5 \times 10^{-6}$ au
8.	Pump pulse noise	N_p	$<2 \times 10^{-6}$ au

Application of Equation 1 predicts a system noise of 4.6×10^{-5} au. Current commercial array detectors exhibit noise of 5×10^{-5} to 2×10^{-4} au.

The noise contributions for the array detector differ significantly from those of the variable and fixed wavelength detectors due to the presence of a dominant electronic readout noise specific to scanning of the array, combined with a relatively low optical throughput reflected in a low value of I_s. These factors result in array readout and shot noise dominance.

Future reduction in array detector noise should focus on reducing the readout noise N_a. Production of JFET switches with junction capacitance <1 picofarad is dependent on technology outside the realm of HPLC and is difficult to forecast. Cooling of the JFET switches and thermal control of the scan circuitry warrants study as a means to reduce array readout noise. It would also be beneficial to increase I_s since array readout noise is a Category II noise source, for which an x-fold increase in I_s yields a full x-fold decrease in noise. This could conceivably be achieved by development and use of higher power lamps (e.g., a 200 watt deuterium lamp, a xenon continuum lamp) and by use of optical designs based on plane grating polychromators having higher optical efficiency than the more simple polychromators based on concave gratings.

Finally, the scan rate of the array should be kept as low as possible within the constraints of limiting the integration time so as not to cause band-broadening of the chromatographic peak and of avoiding saturation of the photodiode. In this manner, one can minimize the amount of array readout noise relative to collection of a given sample current. For example, if one were to scan an array at 100 Hz for 1 second, collecting a signal current I_o at each scan and incurring a readout noise N_a at each scan, and then ensemble averaging the 100 scans, the readout noise fraction will be

$$\frac{100 \, n_a}{\sqrt{100}} \bigg/ 100 \, I_o = \frac{1}{10} \frac{n_a}{I_o}$$

However if one were to scan the array at 1 Hz, for 1 second the signal current collected would be 100 I_o but the readout noise would only be n_a and the readout noise fraction will be only

$$\frac{n_a}{100 \, I_o}$$

To increase the saturation current threshold of the array, allowing one to maximize integration time (minimize scan rate) one can use larger area diodes, a decision which must be made prior to an optical design.

Summary

Optical absorbance detection in HPLC is currently limited by several sources of "non-shot" or "non-optical" noise. As shown in Summary Table V, a state of art variable wavelength UV absorbance detector comes within a factor of two of its 8×10^{-6} au shot noise limit. In this case the dominant noise sources are shot noise and Johnson thermal noise.

An optical design of a fixed wavelength absorbance detector based on a 5 watt discrete line UV lamp should be capable of reducing shot noise to 10^{-6} au. However Category III thermally induced noises due to the temperature coefficient of photodiode quantum efficiency, mobile phase absorbance, flow cell optical throughput and pump pulsations dominate the fixed wavelength shot noise, such that noise significantly below 10^{-5} au has not been achieved to date.

Table V.   Noise Source Contributions in 1984 State of Art Optical Absorbance
Detectors – assumes $\tau=0.5$ sec. For a given detector, each noise
source contribution is normalized to that of its shot noise.

NOISE SOURCE	NOISE CONTRIBUTIONS (Normalized to shot noise)		
	VARIABLE WAVELENGTH	FIXED WAVELENGTH	PHOTODIODE ARRAY
Category I Noise Sources			
1,2. Shot noise	1.0 (7.8×10^{-6} au)	1.0 (1.0×10^{-6} au)	1.0 (2.2×10^{-5} au)
Category II Noise Sources			
3. Johnson thermal noise in readout resistor	1.3	0.2	Not applicable
4. Photodiode dark current noise	0.03	0.004	0.01
5. Photodiode array readout noise	Not applicable	Not applicable	1.8
Category III Noise Sources			
6. Photodiode response tempco noise	0.5	≤ 4	0.2
7. Mobile phase/flow cell tempco noise	≤ 0.6	≤ 5	≤ 0.2
8. Pump pulse noise	≤ 0.3	≤ 2	≤ 0.1
Deviation of Detector from Shot Noise Limit	2-fold	7-fold	2-fold (up to 8-fold on some detectors)

The relatively low throughput of reverse optic/polychromator designs utilized in photodiode array UV absorbance detectors and an electronic readout noise incurred in scanning an array currently limit noise in this detector to approximately 5×10^{-5} au.

The brightest hope for reducing the non-shot and shot noise sources of variable wavelength and diode array detectors would be the development of higher power light sources with emission geometries compatible with micro-flow cell optical collection. Tunable dye lasers with crystal doubling into the UV region would provide the source power and geometry required but are not yet feasible in terms of convenience, cost and perhaps stability. Study of higher power deuterium or xenon lamps appears warranted. Thermoelectric cooling of the readout switches of a photodiode array and thermal control of its scan circuitry also appears warranted for study.

Reducing the non-shot Category III noise sources of the fixed wavelength detector will require attention to improved thermomechanical detector design to reduce noise due to thermal variations imposed on the mobile phase/flow cell and on the sample and reference photodiodes, and continued reduction in HPLC reciprocating pump flow pulsations.

Literature Cited

1. Advances in Chromatography, Vol. 12, Chapter 6, ed. C. Giddings, et al., Marcel Dekker, New York.
2. J. E. Stewart, Appl. Optics, 20, 654 (1981).
3. Applied Optics and Optical Engineering, Volume VIII, Chapter 7, ed. R. R. Shannon and J. C. Wyant, Academic Press, New York, 1980.
4. Y. Talmi, Multichannel Image Detectors, ACS Symposium Series 102, Chapter 6, American Chemical Society, Washington, D.C., 1979.
5. S. R. Abbott and J. Tusa, J. Liq. Chrom., 6, 77 (1983).

RECEIVED May 6, 1985

8

Applications of Laser Fluorimetry to Microcolumn Liquid Chromatography

V. L. McGuffin[1] and R. N. Zare

Department of Chemistry, Stanford University, Stanford, CA 94305

Laser-induced fluorescence (LIF) is examined as a sensitive and selective detector for microcolumn liquid chromatography (LC). This detector employs a cw helium-cadmium laser (325 nm, 5-10 mW) as the excitation source together with a simple filter/ photomultiplier optical system. Several flowcells are evaluated: (1) a flowing droplet; (2) an ensheathed effluent stream; and (3) a fused-silica capillary. The latter is judged to be the most promising for general use. The LIF detector is capable of sensing femtogram amounts of model solutes (coumarin dyes), and the response is linear over more than eight orders of magnitude in concentration. The potential of combining the high separation power of microcolumn LC with the high sensitivity of LIF detection is demonstrated through the analysis of multicomponent mixtures of polynuclear aromatic hydrocarbons and derivatized amino acids.

The identification and quantitation of individual components in complex samples presents a challenging analytical problem that demands continual improvement of existing methodology and instrumentation. This demand has been a primary motivation for the development of high-efficiency separation methods, such as gas (GC), supercritical fluid (SFC), and liquid (LC) chromatographic techniques. In particular, microcolumn liquid chromatography is rapidly gaining popularity for the separation of multicomponent mixtures of nonvolatile compounds. The microcolumns currently under development are of three general types: narrow-bore or microbore packed columns (1-4), semipermeable packed capillaries (5-7), and open tubular capillaries (8-11). Although the theoretical potential and the current state of the art are

[1]Current address: Department of Chemistry, Michigan State University, East Lansing, MI 48824

different for each column type, these microcolumns have
demonstrated several distinct advantages over conventional LC
columns. First, microcolumns are capable of achieving much higher
chromatographic efficiency than their conventional counterparts,
exceeding one million theoretical plates (4,11). Consequently,
microcolumns will improve the separation of very complex samples or
hard-to-resolve solutes. Alternatively, a given separation
efficiency can be attained in a shorter analysis time using
semipermeable packed capillary or open tubular capillary columns
because of their high permeability (12). Another desirable
attribute of microcolumns is their low volumetric flowrates
(1-50 µL/min), which result in significantly reduced consumption of
both sample and solvent. Although the economic and environmental
advantages of reduced solvent consumption are immediately obvious,
other benefits have become apparent with the continued development
and use of microcolumn LC. For example, separations may be
performed using novel mobile and stationary phases that are too
expensive, rare, or toxic to be used with conventional columns
(13). Furthermore, novel detection techniques can be implemented
that are largely incompatible with conventional columns, among
which flame- or plasma-based detectors (14-18) and mass
spectrometry (19,20) are representative examples.

Despite the many advantages of microcolumn LC, there are
several limitations that may ultimately restrict the practical
application of this technique. Long analysis times, ranging from a
few to many hours, are frequently necessary to provide resolutions
of very complex samples. This problem is not characteristic of
microcolumns themselves, but is inherent in all high-efficiency LC
separations because of slow diffusional processes in the condensed
phase. Another limitation of microcolumn LC is the stringent
technological requirements imposed on ancillary chromatographic
equipment such as pumps, sample injectors, connecting hardware, and
detectors. Although significant improvements have been achieved
for many components of the chromatographic system, there has been a
conspicuous lack of sensitive, low-volume detectors. In most
current applications of microcolumn LC, the detectors are merely
miniaturized versions of those employed with conventional columns,
such as UV-absorbance, fluorescence, and electrochemical
detectors. Whereas such devices are adequate at the present stage
of column development, the technological limitations and the
consequences of further miniaturization become immediately
apparent.

In order to achieve the full potential of microcolumn LC, it
is necessary to develop new detection systems that are in
compliance with the rigorous requirements of this analytical
technique. Among the many possibilities, laser-based spectroscopic
detectors appear to be particularly well suited for this
application. The intensity of the laser radiation permits very
high sensitivity to be achieved using those spectroscopic
techniques in which the signal is proportional to source intensity;
for example, fluorescence (21-23), phosphorescence (24-26), light
scattering (27,28), and thermooptic or thermoacoustic measurements
(29-31). Moreover, the highly collimated laser radiation can be
readily focused into flowcells of nanoliter volume, as required for

microcolumn LC, without concomitant loss of radiant power. Other
properties of laser sources, such as narrow spectral bandwidth,
polarization, and temporal characteristics (pulsed or cw), have
been favorably exploited for liquid chromatographic detection as
well (32).

Laser-induced fluorescence (LIF) is one of the simplest and
most promising applications of laser-based detection in liquid
chromatography. Previous investigations with conventional LC
columns have clearly indicated the high sensitivity and selectivity
that can be attained (33-40). Based on these preliminary and
highly promising results, several laboratories have undertaken
concurrently the combination of microcolumn liquid chromatography
and laser fluorimetric detection (41-44). In the present study, we
have developed a sensitive low-volume LIF detector that is
compatible with all of the microcolumns described previously, yet
is simple and reliable to operate. Several flowcells were
constructed and evaluated, including a fused-silica capillary (45),
a miniaturized flowing droplet (33,46) and an ensheathed effluent
stream (21,22,35). The LIF detector was optimized and
characterized with respect to sensitivity, linear dynamic range,
and dead volume. Finally, the outstanding performance of this
detection system was demonstrated through the analysis of
polynuclear aromatic hydrocarbon and derivatized amino acid
samples.

Experimental Section

A schematic diagram of the liquid chromatography system and laser
fluorescence detector is shown in Figure 1. The key components of
this analytical system are described sequentially in what follows:

Liquid Chromatography System. The solvent delivery system was
constructed of two 10-mL stainless-steel syringe pumps (MPLC
Micropump, Brownlee Labs, Santa Clara, CA). By splitting the pump
effluent between the microcolumn and a restricting capillary
(1:20-1:2000), isocratic separations were achieved reproducibly at
column flowrates as low as 0.005 μL/min, and gradient separations
as low as 0.1 μL/min. Samples of 0.5 to 50 nL volume were
introduced by the split injection technique with a 1- L valve
injector (Model ECI4W1., Valco Instruments Co., Inc., Houston,
TX). The injection valve, splitting tee, and microcolumn were
maintained at constant temperature ($\pm 0.3°C$) in a thermostatted
water bath.

Three different types of microcolumns were prepared. Packed
microcolumns were fabricated from fused-silica tubing (Hewlett-
Packard, Avondale, PA) of 0.20 to 0.32 mm inner diameter and 1 to
2 m length. This tubing was packed under moderate pressure
(400 atm) with a slurry of the chromatographic material [Aquapore
RP-300 (10 μm), Brownlee Labs; Micro-Pak SP-18 (3 μm), Varian
Instrument Group, Walnut Creek, CA] in an appropriate solvent (3).
In this manner, narrow-bore packed microcolumns reproducibly
yielded 150,000 or more theoretical plates. Semipermeable packed
capillary columns were prepared by loosely packing a Pyrex glass
tube with an irregular silica adsorbent [LiChrosorb Si-60 (30 μm),

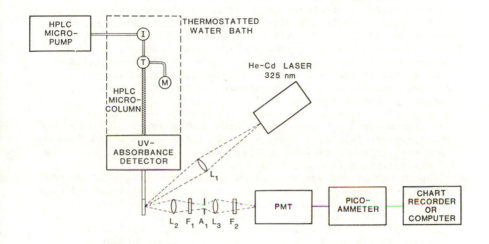

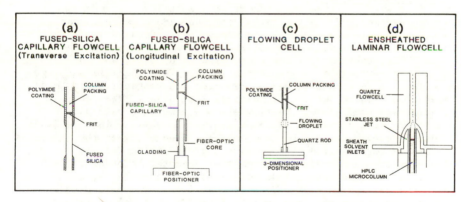

Figure 1.   Schematic diagram of the liquid chromatography
system and laser fluorescence detector with different
flowcells: (a)-(d).   I = injection valve, T = splitting tee,
M = metering valve or restricting capillary, L = lens, F =
filter, A = aperture, PMT = photomultiplier tube.

E. Merck Reagents, Darmstadt, F.R.G.], and subsequently extruding
the packed capillary with a glass-drawing apparatus to 70 μm i.d.
and 25 m length (7). Open tubular capillary columns were
fabricated from fused-silica tubing of 10 to 100 μm i.d. and 5 m
length (Scientific Glass Engineering, Inc., Austin, TX), and were
used without further surface modification. The latter microcolumns
were employed solely to ascertain compatibility with the laser
fluorescence detector and to optimize flowcell design.

A variable-wavelength UV-absorbance detector (Model Uvidec
100-V, Jasco Inc., Tokyo, Japan) was modified to permit "on-column"
detection with packed and open tubular fused-silica microcolumns,
as described previously (3,45,47). The UV-absorbance detector was
placed in series before the laser fluorescence detector (see
Figure 1), and was used for comparisons of sensitivity,
selectivity, and dead volume (44).

Laser Fluorescence Detector. A helium-cadmium laser (Model 4240B,
Liconix, Sunnyvale, CA) was chosen as the excitation source because
of its stability and convenient wavelengths (325 and 442 nm). The
UV laser radiation (325 nm, 5-10 mW cw) was isolated with a
dielectric mirror and was focused on the miniaturized flowcell with
a quartz lens. Sample fluorescence, collected perpendicular to and
coplanar with the excitation beam, was spectrally isolated by
appropriate interference filters and then focused on a
photomultiplier tube (Centronic Model Q 4249 B, Bailey Instruments
Co., Inc., Saddle Brook, NJ). The resulting photocurrent was
amplified with a picoammeter (Model 480, Keithley Instruments,
Inc., Cleveland, OH), and finally was displayed on a stripchart
recorder (Model 585, Linear Instruments Corp., Reno, NV).

Several miniaturized flowcells were evaluated, including a
fused-silica capillary, a suspended flowing droplet, and an
ensheathed effluent stream (Figures la-d). The fused-silica
capillary flowcell was formed by removing the protective polyimide
layer from a short section of fused-silica tubing. When the LIF
detector was used alone, the flowcell was simply an extension of
the column itself (0.20 to 0.33 mm i.d.), thereby eliminating dead
volume from connecting tubes and unions. However, when the
UV-absorbance and LIF detectors were employed in series, it was
necessary to use a capillary of smaller diameter in order to reduce
laminar dispersion between the detectors. In such cases, a
fused-silica capillary of 0.035-0.100 mm i.d. and 1 m length was
attached to the microcolumn outlet with PTFE (Teflon) tubing, and
formed the flowcells for both UV-absorbance and LIF detectors. By
varying the inner diameter, flowcells were constructed with
illuminated volumes from 1 to 100 nL, and corresponding optical
pathlengths of 0.035 to 0.33 mm. Illumination was achieved in
either the transverse (Figure la) or longitudinal (Figure lb)
direction with respect to the flowcell axis. The latter method,
which employed a UV-transmitting optical waveguide inserted
directly into the capillary flowcell, was more easily aligned and
permitted control of the illuminated volume without requiring a
change in capillary diameter. The flowing droplet cell,
illustrated in Figure lc, was a straightforward miniaturization of
the windowless fluorescence flowcell described for conventional

liquid chromatography by Diebold and Zare (33). This cell was
formed by suspending a flowing droplet of the microcolumn effluent
between a fused-silica capillary and a quartz rod of similar outer
diameter. By varying the diameter and spacing of the capillary and
rod, the droplet volume could be adjusted from approximately 50 to
200 nL, with corresponding optical pathlengths of 0.15 to 0.5 mm.
The ensheathed effluent flowcell, shown schematically in Figure 1d,
was both structurally and functionally similar to those described
by Hershberger et al. (35) and Dovichi et al. (21,22). This
flowcell was obtained from a commercial flow cytometer (System 50,
Ortho Diagnostic Systems, Inc., Westwood, MA), and consisted of a
quartz tube of square cross section having a 0.25 mm square central
bore. The effluent was introduced directly from the fused-silica
microcolumn, which extended just beyond a conical stainless-steel
jet at the bottom of the quartz tube. The sheath solvent, which
was of similar composition to the column effluent, was supplied by
a gas-pressurized liquid reservoir through two stainless-steel
tubes perpendicular to and slightly below the column inlet.
Turbulence and mixing of the effluent and sheath streams were not
observed when pulseless, laminar flow conditions were maintained.
The optical pathlength and effective volume of the effluent stream
were controlled by hydrodynamic focusing, by varying the relative
flowrates of the column effluent and the sheath solvent. In this
manner, the pathlength was readily varied between 5 and 20 μm, with
corresponding illuminated volumes of 20 to 300 pL (calculated
values). The large volume of sheath solvent employed, typically
0.5 to 5.0 mL/min, eliminated the advantage gained by microcolumns
in reducing solvent consumption.

Analytical Methodology. An N.B.S. Standard Reference Material (SRM
1647) containing sixteen polynuclear aromatic hydrocarbons was
diluted twenty-fold with methanol/methylene chloride (1:1) prior to
analysis.

Amino acids were derivatized with 1-dimethylaminonaphthalene-
5-sulfonyl chloride (dansyl chloride) according to the procedure of
Tapuhi and coworkers (48-50). A 10^{-3} M stock solution of twenty
common L-amino acids (Sigma Chemical Co., St. Louis, MO) was
prepared by dissolving the carefully weighed standards in 0.1 M
aqueous hydrochloric acid. Aliquots of this solution were
transferred to conical vials, evaporated to dryness, and
redissolved in 500 μL aqueous buffer (0.04 M lithium carbonate, pH
9.5). A 500 μL volume of dansyl chloride solution (5×10^{-3} M in
acetonitrile) was added, and the derivatization was allowed to
proceed in the dark at 35°C for one hour. The reaction was
terminated by the addition of 2% methylamine hydrochloride, and the
derivatized sample was analyzed immediately by microcolumn liquid
chromatography with UV-absorbance and LIF detection.

Organic solvents employed in this investigation were
high-purity, distilled-in-glass grade (Burdick & Jackson
Laboratories, Inc., Muskegon, MI); water was deionized and doubly
distilled in glass (Mega-Pure System, Corning Glass Works, Corning,
NY).

Results and Discussion

Evaluation of Flowcell Performance. There are many important
factors to be considered in the evaluation of flowcell performance
for microcolumn liquid chromatography. Foremost among such
criteria is sensitivity; i.e., the signal-to-noise ratio that can
be obtained for a standard sample injection. If we assume that the
flowcells are uniformly illuminated in the laser beam and that the
fluorescent emission is collected from each cell with the same
efficiency, then the signal intensity should be primarily a
function of the optical pathlength of each flowcell. Accordingly,
the flowing droplet cell, which had the longest pathlength
(0.15-0.5 mm), provided the largest photocurrent, while the
fused-silica capillary (0.05-0.32 mm) and the ensheathed effluent
(0.005-0.02 mm) flowcells yielded proportionately smaller
responses. However, the background noise level also appeared to be
nearly proportional to pathlength, indicating that the predominant
source of noise was fluorescent or scattered (Rayleigh and Raman)
light from the effluent itself, rather than from cell walls.
Specifically, Raman scattering from the solvent seemed to be the
major component of background interference, particularly when the
detected emission wavelength was near the excitation wavelength.
Thus, while the flowing droplet and ensheathed effluent flowcells
were able to eliminate or discriminate against light originating
from cell walls through spatial filtering, this did not translate
directly to an improvement in signal-to-noise ratio. Indeed, this
figure of merit was approximately the same for each of the
flowcells investigated when appropriate spectral filtering was
employed. Consequently, other performance criteria, such as dead
volume, simplicity of operation, and other factors dictated the
preference in flowcell design.
     It is imperative that the volume of the flowcell and
associated connections does not contribute excessively to the
dispersion of the chromatographic peaks. Moreover, it is desirable
to have a single flowcell design that will accommodate microcolumns
of different types and sizes, and will adapt to their different
volumetric requirements. The flowing droplet cell, which could be
varied from 50 to 200 nL, was found to be suitable for packed
microcolumns of 0.5 mm or greater inner diameter, but to suffer
from reduced sensitivity and stability for columns of smaller
bore. On the other hand, the very small volume of the ensheathed
effluent flowcell (20-300 pL) made it suitable for open tubular and
semipermeable packed capillary columns, but somewhat impractical
for columns of larger bore because of reduced sensitivity and
increased consumption of ensheathing solvent. In contrast,
fused-silica capillary cells of varying dimensions could be readily
interchanged to provide a suitable compromise between sensitivity
and dead volume for all of the microcolumns under study.
     Another important consideration is the compatibility of the
flowcell with the normal range of operating conditions in liquid
chromatography. First, the flowcell should be compatible with a
wide variety of solvents for both normal- and reversed-phase
separations. Second, it should readily accommodate variations in
solvent flowrate to permit both high-efficiency and high-speed

applications. Furthermore, for optimal analysis of complex
samples, the flowcell should allow gradients in either solvent
composition or flowrate to be employed. Finally, it is desirable
for the detector cell to be insensitive to temperature and pressure
fluctuations in order to minimize background noise. Although our
investigations encompassed a wide range of operating conditions,
the results can be briefly summarized in the following manner: For
a fixed set of operating conditions (solvent composition, flowrate,
temperature, pressure), the flowing droplet and ensheathed effluent
flowcells could be readily optimized. However, because these cells
have flexible boundaries in the optical region, they were strongly
dependent upon the physical properties of the effluent and could
not readily tolerate changes in operating conditions without
reoptimization. Hence, these cells were not compatible with
gradients in either solvent composition or flowrate. Moreover,
both the flowing droplet and ensheathed effluent flowcells were
sensitive to pressure and flow fluctuations from the pump,
mechanical vibration, and bubble formation. Furthermore, these
cells were essentially destructive; i.e., they caused dilution or
mixing of the column effluent such that another detector could not
readily be employed in series. In contrast, the fused-silica
capillary cell exhibited only nominal dependence upon the
chromatographic conditions.

Because of its operational simplicity, compatibility with many
solvents as well as gradient elution, compatibility with a wide
range of microcolumn types and sizes, and nondestructive nature,
the fused-silica capillary flowcell was judged to be the most
versatile and useful among those investigated.

Characterization of LIF Detector. In a previous study (44), the
LIF detector was characterized with the fused-silica flowcell using
transverse excitation. The limit of detection, measured at a
signal-to-noise ratio of seven (99.9% confidence level) for three
replicate measurements, was determined to be 2.3×10^{-15} g of
coumarin 440 (1.3×10^{-17} moles) injected in a 25 nL sample.
Whereas detection limits as much as an order of magnitude lower
have been reported for laser fluorimetry (22), such results were
obtained under rather rigorous conditions. In contrast, the
detection limit demonstrated in the previous study (44) was
achieved with a relatively low-power laser and simple optical
system using a representative solute analyzed under normal
chromatographic conditions. Hence, this high sensitivity can be
routinely achieved under practical operating conditions.

The fluorescence signal was found to be linearly related to
solute concentration over at least eight orders of magnitude,
extending from the detection limit (5×10^{-10} M) to the solubility
limit (2×10^{-2} M) of coumarin 440 in methanol. It was suggested
that the small diameter of the capillary flowcell effectively
reduced nonlinear behavior in both the absorption and emission
processes, thereby reducing both inner and outer filter effects and
extending the linear response into very high concentrations (44).
This extraordinary linear range facilitates the simultaneous
quantitation of both major and minor components in complex sample
matrices.

There are several important sources of band broadening in the
LIF detector, including both volumetric and temporal
contributions. These contributions were assessed independently in
order to establish the major sources of dispersion and to minimize
their effects (44). The volumetric variance, given by the second
statistical moment of the chromatographic peak, was determined to
be from 0.06 nL^2 to 0.06 μL^2 for capillary flowcells of 0.035 to
0.33 mm diameter, respectively. This variance was predominantly
due to laminar dispersion occurring between the microcolumn frit
and the illuminated region (see Figure 1a). The temporal
dispersion, given by an exponential time constant of 0.7 ms, was
equivalent to 0.0001 nL^2 in variance units. Thus, both volumetric
and temporal variances were sufficiently small for almost any
application in microcolumn liquid chromatography, including both
high-speed and high-efficiency separations.

Applications. Many molecules of environmental and biochemical
significance are inherently fluorescent, among which the
polynuclear aromatic hydrocarbons (PAH) are a representative
example. Although such compounds may occur in very complex sample
matrices, they can be efficiently separated by microcolumn LC and
sensitively detected by laser fluorimetry. An exemplary
chromatogram is shown in Figure 2, wherein a Standard Reference
Material (SRM 1647) containing sixteen polynuclear aromatic
hydrocarbons has been analyzed. This separation was achieved on a
fused-silica microcolumn (0.2 mm i.d., 1.3 m length) containing a
3-μm octadecylsilica packing material (N=150,000) using an
isocratic mobile phase of 92.5% aqueous acetonitrile. For
comparison of sensitivity and selectivity, the chromatograms
obtained with UV-absorbance and laser-induced fluorescence
detection are illustrated in Figure 2. Several important features
of LIF detection become apparent: Whenever fluorescence detection
is applicable, the sensitivity is generally far superior to
UV-absorbance and other common detection methods, frequently by
several orders of magnitude. However, not all solutes of interest
are naturally fluorescent, and very similar molecules may have
widely differing absorption coefficients and quantum yields. This
becomes evident when comparing the relative response of
benzo(g,h,i)perylene and indeno(1,2,3-cd)pyrene, two six-ring
polynuclear aromatic compounds present in equimolar amounts in the
Standard Reference Material. In this case, fluorescence of the
alternate PAH benzoperylene is significantly lower than that of the
non-alternate indenopyrene due to differences in the absorption
coefficient (51) and to quenching by the solvent acetonitrile.
This high degree of selectivity can be advantageous for the
simplification of complex chromatograms and for the reduction of
background interferences.

Because most common UV lasers operate at fixed wavelengths,
the range of application of laser-induced fluorescence may be
somewhat limited. The He-Cd laser provides additional versatility
because two excitation wavelengths are available, 325 nm and
442 nm. This feature is illustrated in Figure 3, taken from the
recent work of Guthrie, Jorgenson, and Dluzneski (41), where a
solvent-refined coal fluid has been analyzed using an open tubular

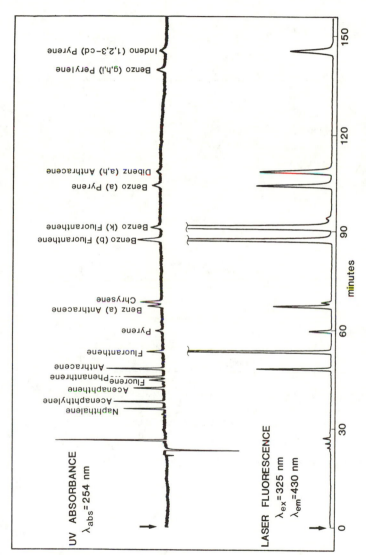

Figure 2. Chromatogram of polynuclear aromatic hydrocarbon standards. Column: fused-silica capillary (0.2 mm i.d., 1.3 m length) packed with Micro-Pak SP-18 (3 μm); Mobile phase: 92.5% aqueous acetonitrile at 1.2 μL/min; Sample injection: 50 nL, 0.2-2.5 ng per component. Reproduced with permission from Ref. 43. Copyright 1984, American Assoc. for the Advancement of Science.

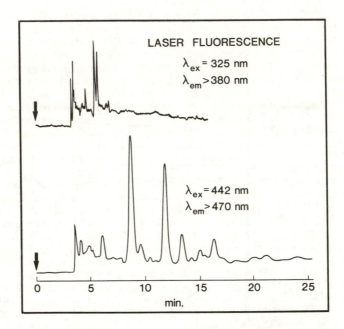

Figure 3. Characterization of a solvent-refined coal sample
by laser fluorimetric detection with dual excitation
wavelengths. Column: open tubular glass capillary
(16 μm i.d., 2.3 m length) with octadecylsilane bonded phase;
Mobile phase: 50% aqueous acetonitrile , 425 psi. Reproduced with
permission from Ref. 41. Copyright 1984, Preston Publications.

capillary column with a chemically bonded octadecylsilane
stationary phase. It is evident that UV excitation at 325 nm
(0.30 mW) is appropriate for the detection of small polynuclear
aromatic hydrocarbons, while visible excitation at 442 nm (4.1 mW)
permits greater sensitivity for PAH compounds of higher molecular
weight. The complementary information contained in these
chromatograms provides a powerful analytical tool for the
characterization of complex coal-derived samples.

Molecules that are not amenable to LIF detection because of
their low absorption cross sections or low fluorescence quantum
yields may be analyzed through a variety of indirect fluorescence
techniques. For example, non-fluorescent solutes may be detected
by the extent to which they enhance (52) or quench (53) the
emission of a fluorescent dye added to the mobile phase.
Alternatively, simple displacement of the dye molecule in the
column effluent by a non-fluorescent solute will result in a
reduction of the fluorescence intensity, which can provide nearly
universal detection capability with adequate sensitivity for many
applications (54). Finally, non-fluorescent molecules may be
rendered detectable through the incorporation of a fluorescent
label that is optimized for the laser excitation wavelength.
Derivatization methods are particularly attractive because they
introduce an additional dimension of chemical and spectroscopic
selectivity that can simplify analyses in a predictable and
reproducible manner.

A wide variety of fluorescent molecular probes have been
demonstrated to be suitable for excitation by the He-Cd laser:
4-bromomethyl-7-methoxycoumarin has been employed for the detection
of carboxylic and phosphoric acids (44,55), 7-chlorocarbonyl-
methoxy-4-methylcoumarin for hydroxyl compounds (42), 7-isothiocya-
nato-4-methylcoumarin for amines and amino acids (55), 7-diazo-4-
methyl-coumarin for a variety of aromatic compounds (55), and
terbium chelate molecules with long fluorescence lifetimes (~ 1 ms)
for protein analysis (56). In this study, we examined the utility
of 1-dimethylaminonaphthalene-5-sulfonyl chloride (dansyl chloride)
as a sensitive and selective reagent for the determination of
biogenic amines and amino acids.

The use of dansyl chloride as a derivatization reagent for
amino acids and peptides is a well established and well accepted
methodology. Nevertheless, improvements in the separation and
detection of these derivatives may have a significant impact on
clinical and biomedical research. The fluorescence excitation and
emission spectra of the dansyl amino acids are shown in Figure 4,
whereupon the laser excitation wavelength (325 nm) and filter
emission wavelength (546 nm, 10 nm FWHM) have been indicated. It
is evident that this fluorescent label is not excited optimally by
the He-Cd laser, yet there appears to be sufficient overlap to
permit sensitive detection. The analysis of twenty common amino
acids derivatives (50-100 pmol each) is demonstrated in Figure 5.
This separation was achieved using a fused-silica microcolumn
containing Aquapore RP-300, a wide-pore octylsilica packing
material of 10-µm nominal diameter. The mobile phase consisted of
an optimized gradient from 5% to 32% 2-propanol in a formic and
acetic acid buffer (pH 3.2). The chromatographic separation shown

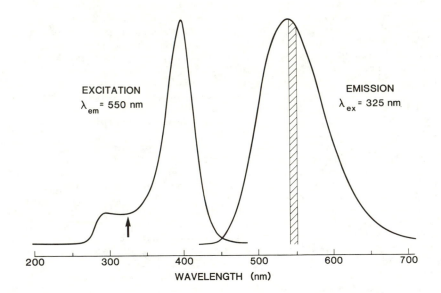

Figure 4.   Fluorescence excitation and emission spectra of dansyl amino acids.

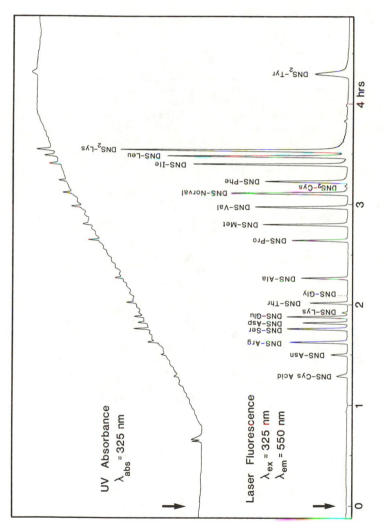

Figure 5. Chromatogram of dansyl amino acid standards.
Column: fused-silica capillary (0.32 mm i.d., 1.3 m length)
packed with Aquapore RP-300 (10 μm); Mobile phase: linear
gradient from 5% to 32% 2-propanol in 0.05 M formic acid/0.06
M acetic acid buffer (pH 3.2) in 200 min, 2.0 μL/min, 55°C;
Sample injection: 130 nL, 50-100 pmol per component.

here required four hours for completion when run at optimum
flowrate (2.0 µL/min), yet it clearly reveals the high resolving
power of microcolumn LC. It is possible and frequently desirable
to compromise separation efficiency to achieve faster analysis
time; a slightly inferior chromatogram, still with baseline
resolution, can be obtained in less than two hours. For purposes
of comparison, the separation was monitored using both
UV-absorbance and laser fluorescence detection. The UV-absorbance
chromatogram (Figure 5, upper trace) shows relatively poor
sensitivity for the dansyl amino acids, even at the optimum
absorbance wavelength. Because this detector is highly sensitive
to changes in absorbance and refractive index of the mobile phase
during gradient elution, the small peaks of interest are
superimposed upon a steeply sloping baseline. In contrast, the
fluorescence chromatogram (lower trace) demonstrates a superior
signal-to-noise ratio with the peaks of interest superimposed upon
an absolutely flat baseline. The high sensitivity of the LIF
detector permits the determination of all dansyl amino acids, as
well as some impurities therein, at the subpicomole level.

Summary

The development of suitable detectors is currently one of the
limiting factors in the routine practical application of
microcolumn liquid chromatography. The laser-induced fluorescence
detector described herein appears to fulfill the rigorous
requirements because of its high sensitivity (a few femtograms of
analyte detected), remarkable linearity (greater than 10^8 dynamic
range), and small detection volume (0.06 nL^2 to 0.06 $µL^2$ for
flowcells of 35 to 330 µm i.d., respectively). Although many
improvements are possible, such refinements would involve a
sacrifice in the inherent simplicity, reliability, and
affordability of the present detection system. This system can
already be applied to a wide variety of interesting analytical
problems, of which the separations of polynuclear aromatic
hydrocarbons and dansyl amino acids are only representative
examples. It would appear that the combination of laser-induced
fluorescence detection and microcolumn liquid chromatography shows
exceptional promise for the analysis of complex samples of
biochemical and environmental origin.

Acknowledgments

We are grateful to Brownlee Labs, Santa Clara, CA, for the loan of
a prototype MPLC Micropump solvent delivery system. We also thank
Raymond Dandeneau of Hewlett-Packard and Ernest Dawes of Scientific
Glass Engineering for providing fused-silica capillary tubing of
non-standard dimensions, and Bojan Petek of the San Francisco Laser
Center for advice on the use of optical fibers.
    This research was supported by the National Institutes of
Health under grant number 9R01 GM29276-06. R. N. Z. gratefully
acknowledges support through the Shell Distinguished Chairs
Program, funded by the Shell Companies Foundation, Inc.

Literature Cited

1. Scott, R. P. W.; Kucera, P. J. Chromatogr. 1979, 169, 51-72.
2. Yang, F. J. J. Chromatogr. 1982, 236, 265-77.
3. Gluckman, J. C.; Hirose, A.; McGuffin, V. L.; Novotny, M. Chromatographia 1983, 17, 303-9.
4. Menet, H. G.; Gareil, P. C.; Rosset, R. H. Anal. Chem. 1984, 56, 1770-3.
5. Tsuda, T.; Novotny, M. Anal. Chem. 1978, 50, 271-5.
6. Hirata, Y.; Novotny, M.; Tsuda, T.; Ishii, D. Anal. Chem. 1979, 51, 1807-9.
7. McGuffin, V. L.; Novotny, M. J. Chromatogr. 1983, 255, 381-93.
8. Tsuda, T.; Hibi, K.; Nakanishi, T.; Takeuchi, T.; Ishii, D. J. Chromatogr. 1978, 158, 227-32.
9. Tsuda, T.; Nakagawa, G. J. Chromatogr. 1983, 268, 369-74.
10. Jorgenson, J. W.; Guthrie, E. J. J. Chromatogr. 1983, 255, 335-48.
11. Kucera, P.; Guiochon, G. J. Chromatogr. 1984, 283, 1-20.
12. Knox, J. H. J. Chromatogr. Sci. 1980, 18, 453-61.
13. McGuffin, V. L. Liq. Chromatogr. Mag. 1984, 2, 282-8.
14. Krejci, M.; Tesarik, K.; Rusek, M.; Pajurek, J. J. Chromatogr. 1981, 218, 167-78.
15. McGuffin, V. L.; Novotny, M. Anal. Chem. 1981, 53, 946-51.
16. McGuffin, V. L.; Novotny, M. Anal. Chem. 1983, 55, 2296-302.
17. Jinno, K.; Tsuchida, H. Anal. Lett. 1982, 15(A5), 427-37.
18. Jinno, K.; Nakanishi, T. J. High Resoln. Chromatogr. & Chromatogr. Commun. 1983, 6, 210-1.
19. Henion, J. D. J. Chromatogr. Sci. 1981, 19, 57-64.
20. Schafer, K. H.; Levson, K. J. Chromatogr. 1981, 206, 245-52.
21. Dovichi, N. J.; Martin, J. C.; Jett, J. H.; Trkula, M.; Keller, R. A. Anal. Chem. 1984, 56, 348-54.
22. Dovichi, N. J.; Martin, J. C.; Jett, J. H.; Keller, R. A. Science 1983, 219, 845-7.
23. Richardson, J. W.; Ando, M. E. Anal. Chem. 1977, 49, 955-9.
24. Cline Love, L. J.; Skrilec, M.; Habarta, J. G. Anal. Chem. 1980, 52, 754-9.
25. Donkerbroek, J. J.; van Eikema Hommes, N. J. R.; Gooijer, C.; Velthorst, N. H.; Frei, R. W. J. Chromatogr. 1983, 255, 581-90.
26. Yamada, S.; Miyoshi, F.; Kano, K.; Ogawa, T. Anal. Chim. Acta 1981, 127, 195-8.
27. Jorgenson, J. W.; Smith, S. L.; Novotny, M. J. Chromatogr. 1977, 142, 233-40.
28. Rogers, L. B.; Stuart, J. D.; Goss, L. P.; Malloy, T. B.; Carreira, L. A. Anal. Chem. 1977, 49, 959-62.
29. Oda, S.; Sawada, T. Anal. Chem. 1981, 53, 471-4.
30. Leach, R. A.; Harris, J. M. J. Chromatogr. 1981, 218, 15-9.
31. Pelletier, M. J.; Thorsheim, H. R.; Harris, J. M. Anal. Chem. 1982, 54, 239-42.
32. Yeung, E. S. Adv. Chromatogr. 1984, 23, 1-63.
33. Diebold, G. J.; Zare, R. N. Science 1977, 196, 1439-41.
34. Folestad, S.; Johnson, L.; Joseffson, B.; Galle, B. Anal. Chem. 1982, 54, 925-9.

35. Hershberger, L. W.; Callis, J. B.; Christian, G. D. Anal.
    Chem. 1979, 51, 1444-6.
36. Furuta, N.; Otsuki, A. Anal. Chem. 1983, 55, 2407-13.
37. Sepaniak, M. J.; Yeung, E. S. Anal. Chem. 1977, 49, 1554-6.
38. Huff, P. B.; Tromberg, B. J.; Sepaniak, M. J. Anal. Chem.
    1982, 54, 946-50.
39. Yeung, E. S.; Sepaniak, M. J. Anal. Chem. 1980, 52,
    1465A-81A.
40. Green, R. B. Anal. Chem. 1983, 55, 20A-32A.
41. Guthrie, E. J.; Jorgenson, J. W.; Dluzneski, P. R.   J.
    Chromatogr. Sci. 1984, 22, 171-176.
42. Gluckman, J.; Shelly, D.; Novotny, M. J. Chromatogr. in
    press.
43. Zare, R. N. Science 1984, 226, 298-303.
44. McGuffin, V. L.; Zare, R. N. Appl. Spectrosc., in press.
45. Yang, F. J. J. High Resoln. Chromatogr. & Chromatogr.
    Commun. 1981, 4, 83-5.
46. Voigtman, E.; Jurgensen, A.; Winefordner, J. D. Anal. Chem.
    1981, 53, 1921-3.
47. McGuffin, V. L.; Novotny, M. Anal. Chem. 1983, 55, 580-3.
48. Tapuhi, Y.; Miller, N.; Karger, B. L. J. Chromatogr. 1981,
    205, 325-37.
49. Tapuhi, Y.; Schmidt, D. E.; Lindner, W.; Karger, B. L. Anal.
    Biochem. 1981, 115, 123-9.
50. De Jong, C.; Hughes, G. J.; van Wieringen, E.; Wilson, K. J.
    J. Chromatogr. 1982, 241, 345-59.
51. Peaden, P. A.; Lee, M. L.; Hirata, Y.; Novotny, M. Anal.
    Chem. 1980, 52, 2268-71.
52. Asmus, P. A.; Jorgenson, J. W.; Novotny, M. J. Chromatogr.
    1976, 126, 317-25.
53. Haugen, G. R.; Richardson, J. H.; Clarkson, J. E.; Hieftje,
    G. M. Proceedings of New Concepts Symposium and Workshop on
    Detection and Identification of Explosives, Reston, VA, 1978.
54. Mho, S.; Yeung, E. S. Iowa State University, Ames, IA,
    private communication.
55. Gassmann, E.; Kuo, J. E.; McGuffin, V. L.; Zare, R. N.
    Stanford University, Stanford, CA, unpublished research.
56. Kuo, J. E.; Milby, K. H.; Hinsberg, W. D.; Poole, P. R.;
    McGuffin, V. L.; Zare, R. N. Clin. Chem. 1985, 31, 50-3.

RECEIVED October 4, 1985

Recent Advances in New and Potentially Novel Detectors in High-Performance Liquid Chromatography and Flow Injection Analysis

Ira S. Krull

Department of Chemistry, The Barnett Institute, Northeastern University, Boston, MA 02115

A review is provided of some of the latest advances in new and potentially novel, selective detectors for high performance liquid chromatography and flow injection analysis. A number of very useful and practical element selective detectors are covered, as these have already been interfaced with both HPLC and/or FIA for trace metal analysis and speciation. Some approaches to metal speciation discussed here include: HPLC-inductively coupled plasma emission, HPLC-direct current plasma emission, and HPLC-microwave induced plasma emission spectroscopy. Most of the remaining detection devices and approaches covered utilize light as part of the overall detection process. Usually, a distinct derivative of the starting analyte is generated, and that new derivative is then detected in a variety of ways. These include: HPLC-photoionization detection, HPLC-photoelectrochemical detection, HPLC-photoconductivity detection, and HPLC-photolysis-electrochemical detection. Mechanisms, instrumentation, details of interfacing with HPLC, detector operations, as well as specific applications for each HPLC-detector case are presented and discussed. Finally, some suggestions are provided for possible future developments and advances in detection methods and instrumentation for both HPLC and FIA.

For at least the past twenty years or more, tremendous progress has been realized in the general field of high performance liquid chromatography (HPLC). Significant progress has been reported in many areas of instrumentation, interfacing, automation, computer/microprocessor control, design, and even robotics control. It has become very clear that HPLC is a dominant analytical area, whose growth continues today. Though chromatographic separations have progressed very rapidly over the past two decades in high performance liquid chromatography, to the current situation where proteins, enzymes, polynucleotides, and much smaller molecules can now be readily resolved and recovered, the

0097–6156/86/0297–0137$08.50/0
© 1986 American Chemical Society

final detection of analytes has progressed somewhat more slowly (1-14).
Any number of support materials, bonded phases, ion exchange packings,
optically active phases, normal bonded phases, etc. have been made,
and these were evaluated for the separation of inorganic and organic
species during at least the past twenty years. Significant advances
have also been made in the uniformity of particle sizes, narrow particle diameters, uniform particle distributions, and related parameters
of importance related to the stationary phase. A large number of mobile phases have also been developed for HPLC separations, modifiers
for these mobile phases, metal additives for the mobile phase, binary
and ternary solvent mixtures, gradient elution, and combinations there-
of.

     Initially, detection involved the refractive index (RI) approach,
and though this crude method is still used for polymer and preparative
separations, it has fallen by the wayside in just about all low-level
analytical determinations. Ultraviolet-visible (UV-VIS) detection also developed quite rapidly from the inception of HPLC, and today there
are commercially available fixed wavelength, variable wavelength,
multiwavelength, linear diode array spectrometers, fast scanning
spectrometers, and other approaches, all of which are general, non-
selective methods of analyte identification. In most HPLC labs today,
there are a number of UV-VIS detectors, and perhaps one fluorescence
(FL) and one electrochemical (EC) detector for HPLC. FL monitors are
quite useful, in that they are very sensitive, routinely operate at
the trace and ultratrace levels, and are relatively selective in that
few compounds exhibit inherent or native fluorescence. However, neither
UV-VIS nor FL detectors really provide any compound identification,
other than the presence of a chromophore and/or fluorophore. EC
detectors have appeared more recently, but they have caught on very
quickly, and appear to offer significant promise both with regard to
selectivity and sensitivity of analysis. Especially with dual electrode approaches, already commercially available from several firms,
EC is able to provide good to excellent analyte identification or
selectivity, together with very good sensitivity and detection limits.
It would appear that LCEC will continue to develop quite rapidly, especially with regard to other applications, both organic and inorganic.
But, it will always be limited to those compounds that have inherent
electrochemical properties. Of course, all detection methods become
suitable for any analyte when off-line or on-line derivatizations
are employed, but there are some serious disadvantages in using homogeneous, solution derivatizations in HPLC (15-17). Ideally, in the
future, all derivatizations will be performed on-line, pre- or post-
column, using reagents that do not get mixed with the HPLC eluent via
two flowing streams. Solid phase reagents in HPLC would avoid the need
for extra pumps, mixing chambers, reaction chambers, dead volume, etc.

     Perhaps UV-VIS, FL, or EC detectors can all be considered some-
what selective, in that they respond only to those compounds that are
chromophoric, fluorophoric, and/or electrophoric. None of them respond
to all compounds passing through, as with the RI, but other than for
wavelength ratioing or dual electrode response ratioing, they cannot
provide true analyte identification or detector selectivity for a
particular functional group, element, or size of molecule. They are
all extremely sensitive, and indeed any detector today that cannot
reach into the low parts-per-billion (ppb) range doesn't receive much
attention in HPLC. The next generation of detectors will have to be

more sensitive, <u>and</u> to provide more selectivity than is now possible with existing detectors. That is, such newer detectors will have to be able to discriminate one analyte peak from the next, and to provide some relatively unambiguous identification or characteristic for each analyte different from others present. We want the detector to see very low levels, and also to tell us something about what it is seeing at such low levels. Though the mass spectrometer (MS) provides true analyte selectivity and identification, it is not about to become a very widely used or easily employed detector in most HPLC situations/ applications. Surely HPLC-MS is here to stay, it is developing quite rapidly, but it may never become as readily used and applied as HPLC-UV or LCEC are already.

The next generation of detectors should tell us something about an analyte or class of analytes that it detects, and/or it should respond to very select compounds, even in the presence of many other similar eluting compounds/analytes. The ultimate selective detector would respond to only one particular compound, but it would not be very practical or economical, since it could not also be used for any other compounds. Ideally, a selective detector should, at times, be selective for one class of analytes, and under other operating conditions, selective for another class, but always provide information for each class separately. The word selective is interpreted incorrectly; it may not mean that the detector responds to a unique compound or compounds. Does it mean that the detector responds to many compounds, but does so differently with different information? If it responds to a very small number of compounds, it is much more selective in its response, but less useful for other classes. If it responds to many classes, it is much less selective, but more useful and practical, and could still provide information that adds to the selectivity of the final response. In general, the term selective has come to mean response to only certain classes of compounds, and the term general means that a detector responds to everything that passes through that detector.

We are going to consider in this chapter a very select group of new and potentially novel detectors, all of which provide for greater selectivity in their responses than UV-VIS, FL, and/or EC. Our choice of which detectors to discuss and consider here has been based, in part, on how we perceive the current trends in detector development and application. It has also been based on our own biased and personal interests, as dictated by our own recent research and development results with these very same detectors.

The requirements for a potentially novel, selective detector in HPLC are several, including: 1) high sensitivity for many classes of analytes, perhaps into the low ppb range or below; 2) high degree of selectivity in responses as dictated by many experimental variables; 3) compatibility with all important HPLC mobile phases; 4) qualitative and quantitative information about a particular analyte that differentiates its response from all or almost all others eluting nearby; 5) reproducibility of response for repeated injections; 6) high degree of accuracy and precision in quantitative determinations; and 7) an ease of interfacing with commercial HPLC instrumentation, fittings, and equipment. Some of these requirements are met by many new detectors, but most new detectors do not meet all of these suggested needs. Some other, perhaps less important, but still practical requirements are that the detector be easy to use, modular in design, inexpensive,

facile to turn on and operate on a daily basis, easy to maintain and
repair, immune to changes in flow rates and back pressures in the HPLC
system, and compatible with multiple detection (series or parallel)
approaches. The detector should also contribute relatively little to
overall band broadening of the HPLC peak, it should be almost unnoticed
in the final peak shape produced, and it should be nondestructive by
returning most or all of the originally injected analyte to the analyst for further studies or determinations. Table I summarizes most
of these ideal requirements for our new and potentially novel,
selective detector for HPLC or FIA.

Table I. Ideal Requirements for New and Potentially Novel, Selective
                  Detectors in HPLC or FIA

Suggested Detector Requirements	Reasons for Suggested Requirements
high sensitivity and low detection limits	useful for trace and ultratrace analyses
high degree of selectivity	improved analyte identification and qualitative confirmation
compatible with important HPLC mobile phases	can be used with current separations does not require new methods
qualitative and quantitative information about analytes	improved analyte identification, improved quantitative analyses
reproducibility of responses in peak height and peak area	improved accuracy and precision in quantitative determinations
high degree of accuracy and pre-cision in quantitative analyses	improved overall usefulness of final analysis
ease of interfacing with current HPLC instrumentation	widespread acceptance by community-at-large, greater usage and applications, less expenses
easy-to-use, operate, maintain, and/or repair	more up-time, better acceptance by analysts, more utility, reduced expenses, faster sample turnout
modular in design	may be moved from lab-to-lab, easy to transport, interface, arrange
immune to changes in flow rates, back pressure of column	greater qualitative and quantita-tive reproducibility, more accurate and precise results
compatible with multiple detection	useful with other detectors at the same time to improve analyte identification/confirmation
little additional band spreading	little loss of chromatographic resolution, efficiency, shape
nondestructive to analyte	recovery of injected analyte, more studies on same material possible
unreactive and unresponsive to HPLC mobile phases	low background noise levels, no false positives due to solvent, no large solvent front present

We discuss here a select group of selective detectors, these being:
1) element selective detectors (ESD) that respond selectively to one
or more element at the same or different times, including atomic
absorption spectrometers (AAS), flame (FAA) or graphite furnace (GFAA),
inductively coupled plasma (ICP) emission spectrometers, direct current
plasma (DCP), and only incidentally microwave induced plasma (MIP) or
atomic fluorescence spectrometer (AFS); 2) photoionization detector
(PID) based on a photochemical ionization of the analyte molecule
leading to formation and eventual collection/detection of ions and
electrons; 3) a post-column, photolytic derivatization in LCEC or
HPLC-photolysis-EC (HPLC-hv-EC) that uses a discrete photolytic
degradation of the original analyte as the basic detection principle;
4) a new photoelectrochemical detector (PED) that again uses light
energy as an on-line, post-column derivatization step, but now to
generate a short-lived, possibly excited species that has novel EC
properties; and finally, 5) a possibly novel electrogenerated luminescence detector (ELD) that uses electrochemistry to generate new
species that combine, self-annihilate, and then together luminesce,
this then being detected by a suitable photomultiplier tube or mono-
chromator outside of the ELD cell. Some of these newer approaches to
detection in HPLC or FIA are still in development stages, indeed
perhaps most of them fit that description, but still others have already been fully optimized and remain to be more fully applied to
actual samples and specific applications.

Element Selective Detectors for HPLC and FIA

Trace inorganic analysis and speciation have become very popular and
accepted only within the past decade or thereabouts, especially in
view of the late realization on the part of many environmentalists
and/or toxicologists that different metal species may have different
biological/toxicological responses or effects in animals and man (18-
29). Though a variety of approaches have already been developed to
perform inorganic, especially metal, speciation; including anodic
stripping voltammetry, differential pulse voltammetry and amperometry,
double differential pulse amperometry, differential pulse polarography,
etc., many of these suffer from the lack of an initial separation or
chromatographic step before the selective detection step. Thus, of
late, most reports deal with the interfacing of either gas chromatography (GC) or high performance liquid chromatography (HPLC) with
some type of element selective detection. Prior to the current widespread popularity of various plasma induced emission techniques, this
involved both flame and graphite furnace atomic absorption spectrosco-
pies (FAA/GFAA). This area has most recently been adequately reviewed
both by Schwedt and by Jewett and Brinckman (25, 26). It has long been
our belief that true element selective detection in HPLC can only be
done using some type of atomic absorption or emission spectroscopy
(AAS or AES). Others contend that element selective detection can also
be realized by various types of post-column derivatization, chelation,
complexation, etc., as well as by electrochemical approaches (30-34).
Nevertheless, perhaps some metal and nonmetal speciation studies
today involve some type of GC/HPLC-element selective detection using
atomic absorption or emission spectroscopy. Though FAA and GFAA have
been used for quite some time now, both of these hyphenated methods
in HPLC suffer some serious disadvantages and practical limitations

(35-52). HPLC-FAA approaches generally have impractical detection limits for many real world applications, although sample preconcentration methods can often get around such limitations. Of all element selective detectors for HPLC, perhaps FAA suffers from having the highest minimum detection limits (MDLs). This severely limits its usefulness, except with those samples that already contain very high levels of metal species, and/or those where analyte (metal species) preconcentration is an easy task before HPLC-FAA. HPLC-GFAA, on the other hand, offers perhaps the best MDLs of all element selective detection approaches, at times reaching into the low ppb or high parts-per-trillion (ppt) ranges. This is due to the inherent analyte preconcentration step built into all GFAA methods by the use of a graphite furnace that allows for multiple sample introductions with desolvation prior to atomization-detection of all preconcentrated metal species.

However, a serious disadvantage to HPLC-GFAA is the fact that some type of merry-go-round or automated sample collector and injector must be interfaced between the end of the HPLC column and furnace of the GFAA unit (49-52). Brinckman's group at the U.S. National Bureau of Standards has best pioneered and evaluated the HPLC-GFAA instrumentation, methods, and final applications. Though extremely useful and practical for many environmental samples, the overall system requires a modest investment of time, energy, money, and total instrumentation before it is routinely operative. It also produces final HPLC-GFAA chromatograms that are in histogram fashion, because of the nature of the AA data output, and thus it is not generally possible to obtain a continuous, smooth chromatogram for this system. The HPLC portion of the system must be finely tuned so that there are no overlapping peaks due to the same element containing moieties/analytes, otherwise the GFAA detection step might not be able to fully resolve or separate these for final quantitative measurements. Though quite practical and useful, relatively few groups today use HPLC-GFAA approaches, perhaps because of the non-continuous nature of its data.

Most literature reports today seem to utilize some type of plasma emission detection, be this inductively coupled plasma (ICP) or direct current plasma (DCP). Though microwave induced plasma (MIP) emission spectrometry has been used for at least twenty years in GC, it does not appear at this writing to be very compatible with most conventional HPLC flow rate requirements (20, 21). Though there are some reports using the MIP as a low flow-rate (microbore) HPLC detector, these appear to offer little promise for more widespread acceptance and practical applications. We have recently attempted some HPLC-MIP work, but at any flow rate over 100ul/min, depending on the power of the MIP, the plasma immediately becomes extinguished. Some other groups, notably that of Caruso in Cincinnati, seem to have had more success with a high powered MIP in combination with low flow-rate microbore HPLC. Much more work seems to have been done in HPLC-ICP than in HPLC-DCP (18-22, 53, 54). Though these are very expensive detectors by conventional HPLC-detector standards, they appear to offer some very significant advantages for true element selective detection and overall inorganic, esp. metal, speciation today. It is not yet clear which of these two plasmas is more advantageous in HPLC, though surely more publications appear to use the ICP. This could be simply because more researchers have purchased the ICP since it is commercially available from many more suppliers than the DCP. Only one commercial DCP is available today, that being from Smith, Kline, and Beckman.

Their Spectraspan VI that appeared on the market in 1984 appears to be
an ideal element selective detector for HPLC, though very few papers
have thus far utilized that specific model in HPLC–DCP studies. It is
arguable that MDLs via HPLC–DCP must, by the very nature of the DCP,
always be inferior to the HPLC–ICP system. We have not found this
always to be the case, and indeed in a very recent HPLC–DCP study of
various chromium containing species, the HPLC–DCP MDLs appeared somewhat better (lower) than those reported by HPLC–ICP methods (22). We
have already suggested some possible explanations for these observations, though nobody has yet made a full, direct comparison of HPLC–
DCP vs HPLC–ICP MDLs for even a few metal species.
    There are still some significant problems in both HPLC–ICP and
HPLC–DCP interfacing for practical applications where low MDLs are
generally needed. But for a few miraculous reports using HPLC–ICP that
provided MDLs at or below the published MDLs for direct–ICP work, we
and others have come to realize that the basic nebulizers developed
years ago for direct–ICP are less–than–ideal for any HPLC–ICP interfacing today. Such nebulizers, aside from being inherently inefficient,
were never designed from the start for the constantly changing concentrations of analytes eluting from an HPLC column. Even with the ultrasonic nebulizers that provide for much more efficient sample introduction into the ICP, the final MDLs via HPLC–ICP with this particular
nebulizer are still less than adequate for trace inorganic speciation
requirements today. Many have suggested that this is indeed the final
solution to the HPLC–ICP MDL problem, but we and others are less than
convinced based on the evidence at hand. What then is the real answer
to the major obstacle to more widespread acceptance and utility of
the ICP/DCP systems for HPLC applications, or is there more than one
single answer possible? We tend to believe that there are several
possible solutions to this problem, and these could be: 1) electrothermal or graphite furnace/cup sample vaporization in HPLC–ICP; 2)
post–column hydride derivatization in HPLC–ICP or what is termed
HPLC–HY–ICP; 3) post–column chemical chelation or complexation of
metals in HPLC–ICP; and 4) use of the Vestal thermospray interface for
HPLC–ICP/DCP and related approaches. It would be possible to spend the
rest of this chapter just discussing these four possible approaches
for improved HPLC–ICP/DCP interfacing, but we will attempt to do so
in less than one or two paragraphs.
    The ICP has some significant advantages for trace metal detection
in comparison with both FAA and GFAA, especially with regard to multi-
element capabilities and elimination of most matrix effects. Perhaps
for these very reasons, as well as its current popularity, a great
amount of work appears today in the literature attempting to interface
both GC and HPLC with ICP (18–21, 25, 55–57). Most of this work has
involved conventional ICP nebulizer interfaces, and only within the
past few years have serious alternative approaches for HPLC–ICP interfacing been evaluated for improved MDLs. Electrothermal vaporization
for improved sample introduction into direct–ICP systems has been
described, and already some investigators are using HPLC–electrothermal vaporization–ICP approaches for true metal speciation (78–84).
However, virtually all of these attempts to use electrothermal or
heated tantalum ribbon sample introduction methods for HPLC–ICP coupling are non–continuous, operate in a step–wise manner, require collection of HPLC fractions, and cannot function in a true on–line
mode. At times, depending on the particular analyte species being

studied, these methods also require simultaneous UV detection of the
HPLC eluents, followed by collection of those fractions of interest
for manual or automated introduction into the ICP. These approaches
are reminiscent of the HPLC–GFAA methods, and though producing semi-
continuous HPLC–ICP element specific chromatograms, they are far from
ideal or routine in operations. If the HPLC eluent can be continuously
introduced into a heated graphite rod or tube, with continuous preconcentration and removal of solvent followed by continuous introduction of the analyte alone, this might be a practical approach.

Post-column, on-line, continuous hydride derivatization after
HPLC separation, with continuous introduction of the now-formed metal
hydrides into the ICP appears a truly practical and on-line approach
for various applications (28, 54, 85–87). Figure 1 illustrates in
schematic fashion the instrumental arrangement for performing on-line,
continuous, post-column, real-time hydride derivatization in HPLC–
Hydride Generation-ICP approaches (28). We have now successfully used
this system and approach for the trace analysis and speciation of
arsenic in drinking well waters in the U.S., as well as for the determination of methylated organotins in liquid and solid food samples for
the U.S. Food & Drug Administration (FDA)(54). Figure 2 illustrates a
typical HPLC–HY–DCP chromatogram for three methylated organotins, now
using a paired-ion, reversed phase HPLC separation. MDLs, calibration
plots, linearities, analyte selectivity, and related analytical parameters of importance for true trace analysis and speciation of these
metals capable of forming hydrides, all appear adequate for additional
applications. We have already used HPLC–Cold Vapor-ICP approaches for
organomercury speciation, analogous to what others have done for HPLC–
CV–GFAA of these same mercury species (88). These newer approaches for
metal speciation also appear quite adequate and appropriate for real
applications in the future. Hydride or cold vapor generation techniques
as a separate post-column derivatization step for select elements thus
appears to be a quite practical and viable overall hyphenated method
in HPLC–ICP/DCP for future end uses. Detection limits are presumably
improved over conventional HPLC–ICP/DCP in the absence of this post-
column reaction by several orders of magnitude in all cases studied,
which thus makes the HPLC–HY–ICP/DCP approach practical today for
at least those metals capable of forming hydrides or a cold vapor.

However, that still leaves a very large number of elements in-
capable of forming volatile hydrides, for which some alternative,
practical, on-line approach remains to be developed for future HPLC–
ICP/DCP interfacing. Post-column metal/element chelation or complexation has been used somewhat in the past for direct-ICP/DCP sample
introduction, but most of these results have been less-than-promising
(89–93). We have also studied this approach, with off-line derivatization of various metals into known, presumably volatile derivatives
that have been used in GC analysis. Use of such pre-formed chelates
or complexes in direct-ICP analysis, using conventional cross-flow
nebulizers, has not yielded any significant enhancements in final
sensitivities or MDLs. Even using organic solvents for introduction
of these metal chelates has not yet yielded significant lowering of
the MDLs in comparison with conventional metal introduction into
direct-FAA or direct-ICP systems (94–96). It does not now seem likely
that either volatile metal derivatives and/or the use of more volatile
organic solvents, such as methyl isobutyl ketone (MIBK), will have
pronounced, positive effects on final FAA or ICP/DCP sensitivities/MDLs.

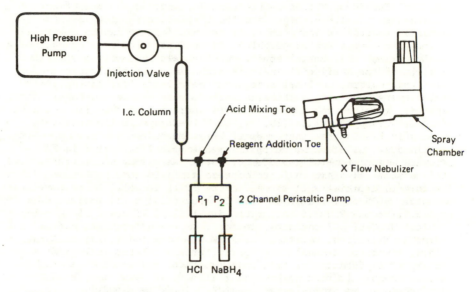

Figure 1. Schematic diagram of the total HPLC-HY-ICP instrumentation. Reproduced with permission from Ref. 28. Copyright 1984,
Marcel Dekker.

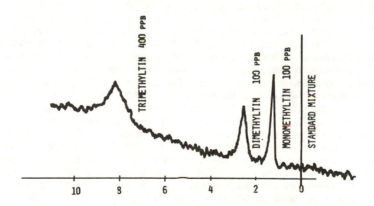

Figure 2. HPLC-HY-DCP chromatogram of mono-, di-, and trimethyltin;
column: 25 cm x 4.1 mm i.d., 10um, PRP-1; mobile phase: 0.003M PIC
B-6/0.003M KF/0.02N H_2SO_4/2.5% HOAc at 2.5 ml/min flow rate; 200ul
loop injection; chart speed 0.5 inch/min; full scale 5 mV. Reproduced with permission from Ref. 54. Copyright 1985, Williams & Wilkins.

One other possible approach for overall improved HPLC-ICP/DCP performance related to lowered MDLs, has to do with the use of a heated spray chamber of even a Vestal-type thermospray introduction system (97-104). The idea of these approaches is to produce a much finer aerosol-mist or heated spray from the HPLC eluent, which is then partially or fully desolvated prior to entering the flame of the FAA or the plume region of the ICP/DCP. In some ways, this is an extension of the current ultrasonic nebulizer, which also produces a very fine mist that is more efficiently transported into the plasma plume area for final detection. Though some work has already been done using a heated aerosol or spray chamber, and much work has been done by Browner's group on the effect of aerosol droplet size as these effect final detection limits, overall results-to-date do not appear very promising for significantly reduced MDLs. Much less has been done on the possible use of a heated spray chamber/nebulizer for HPLC-FAA or HPLC-ICP/DCP, though a great deal of work has already been done in HPLC-mass spectrometry with such interfaces (101-106). We would expect these newer approaches of sample introduction for HPLC-MS to have some possible applications to HPLC-ICP/DCP, and efforts are encouraged to explore such newer interfacing techniques (107).

In closing this section, it should be emphasized that element selective detection in HPLC is an area now undergoing very rapid exploration and development, and significant improvements in HPLC-ICP/DCP detection limits with resultant practical applications should be realized in the near future. If a then relatively inexpensive element selective spectroscopic detector could be developed for conventional or microbore HPLC, with a successful interface providing low MDLs, this could result in a more widespread acceptance and application of both the overall instrumentation and final techniques.

The Photoionization Detector in HPLC or FIA

For many years, the photoionization detector (PID) has been studied and applied in GC, and it is recognized as both a very sensitive and selective approach for trace analysis (108-115). For about the past decade, various people have attempted to interface this same PID with HPLC separations, in order to provide a final HPLC-PID system compatible with reversed and normal phase HPLC solvents (116-119). Most recently, Driscoll et al. have described the successful interfacing of reversed and normal phase HPLC with a high temperature PID (PI-52) via a heated oven that effectively vaporized all of the HPLC effluent prior to its introduction into the PID unit (116). In some ways, this is similar to the thermospray interface for HPLC-MS, as already discussed above. Figure 3 illustrates the overall instrumentation schematic for this particular HPLC-PID approach, wherein all or a fraction of the HPLC effluent can be introduced into the high temperature GC oven. Using this approach, the HPLC effluent is 100% vaporized within the oven, as a function of temperature and residence time therein, dependent on total flow rate entering the oven, and this vaporized mixture of mobile phase solvents and analyte molecules is then passed through to the final PID. The overall approach has been optimized with regard to carrier gas, lamp energies, mobile phase compatibility, flow rate compatibility, and final MDLs. Calibration plots for typical analytes have been over several orders of magnitude starting from the detection limit range. Under the best of

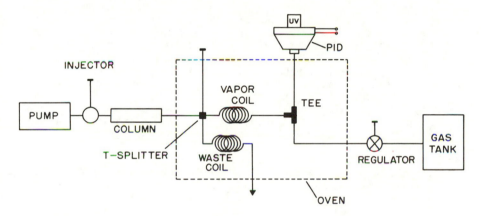

Figure 3. Schematic diagram of the HPLC–PID apparatus in operation. Reproduced with permission from Ref. 116. Copyright 1984, Elsevier Science.

conditions yet realized, and for the compounds showing the greatest
sensitivity, MDLs can be in the low ppb range (5-50 ppm) (116). A wide
variety of classes of organics have already been studied by this HPLC-
PID interface, many of which are summarized in Table II, together with
MDL (ng) and lamp energy (eV) used for those particular compounds.
Typical HPLC-PID chromatograms are illustrated in Figures 4-5, one for
a normal phase separation of N-substituted anilines, the second for
a reversed phase separation of these very same anilines, now eluting
in the exact opposite (reversed) order.

Work is continuing in our labs on improvements in the overall
HPLC-PID interface, the use of volatile organic and inorganic buffers
for reversed phase separations of bioorganics, improvements in MDLs
and sensitivities, and extension of HPLC-PID applications to other
classes of organics and inorganics. In view of the large number of
classes of compounds suitable in GC-PID, often at the ng detection
limits, it is hoped that most of these will be suitable candidates
in HPLC-PID as well. This could suggest the eventual trace determination of amines and amino acids, perhaps peptides and proteins, without
any prior, off-line or on-line derivatization step before detection by
the PID. If this proves feasible, it would be the only sensitive and
selective detector available today in HPLC that could detect such
underivatized materials. This may also prove useful in the eventual
detection of peptides, polypeptides, and perhaps even proteins and
enzymes, which could offer significant advantages over current methods.

The PID works on the principle of ionizing an electron from the
starting compound, usually this is a non-bonding electron on nitrogen,
sulfur, oxygen, or a halogen, though it could be an aromatic Pi-elec-
tron on benzene, and the resultant organic cation and electron are
then captured by the electrodes within the PID to produce a resultant,
overall current. The number of cations and electrons generated per g
or millimole of analyte entering the PID then determines that compound's
sensitivity and final MDLs. This is a function of the ease of photoionization or ionization by photochemical means for that particular
compound, which electrons are available for photoionization, and the
energy and intensity of the light striking these molecules. Thus,
photoionization is really akin to the flame ionization process, but
instead of using a flame, the PID uses light energy to ionize the
original analyte molecules. Operating the PID at a high temperature of
about 275°C ensures that none of the analyte or mobile phase molecules
will recondense on the window of the PID. Use of methanol/water mobile
phases also maintains a very clean window in the PID, so that reproducibility of response is generally quite high in HPLC-PID operations,
even for extended periods of time. Although there is one commercial
system now on the market (HNU Systems, Inc., Newton Highlands, Mass.),
it has only been available for about one year, and it has not yet been
widely applied to real samples. Such work is now in progress, and we
hope to report further results with this very HPLC-PID system in the
near future (120). Not all normal or reversed phase solvents will
prove compatible with the PID requirements, especially those solvents
that have any photoionization propensity or properties. Virtually all
alcohols and water are fully compatible with the PID, but solvents
such as acetone or acetonitrile, and others, are not. This really has
to do with the ionization potentials of these solvents and the IP of
the lamp being used. Those solvents having IPs lower than the lamp
energy will produce very high background noise levels, making them
incompatible with routine, continuous HPLC-PID operation.

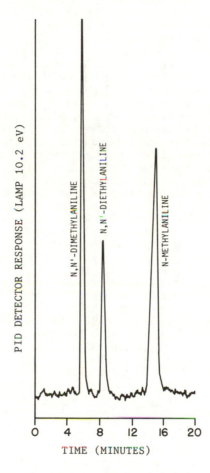

Figure 4. The HPLC-PID chromatogram of N-substituted anilines on a Lichrosorb Si60, 25 cm x 4.6 mm i.d. column. Mobile phase: 3% IPA/ hexane, flow rate 0.46 ml/min, split ratio 3:7 (PID:waste), 10.2 eV lamp, interface at 230°C, PID at 290°C, attenuation setting 1 x 10^{-12} aufs. Reproduced with permission from Ref. 116. Copyright 1984, Elsevier Science.

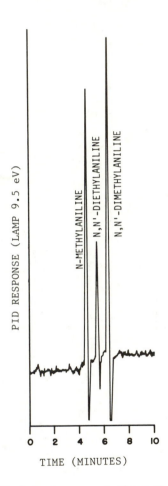

Figure 5. The HPLC-PID chromatogram of three N-substituted anilines:
N-methyl, N,N'-diethylaniline, and N,N'-dimethylaniline. Column:
C_8, 10um, 25 cm x 4.6 mm i.d.; mobile phase: HOH/ACN (25/75, v/v),
flow rate 0.8 ml/min, split ratio 7:3 (PID:waste), 9.5 eV lamp,
interface oven 240°C, PID at 290°C, attenuation setting 2 x 10^{-10}
aufs. Reproduced with permission from Ref. 116. Copyright 1984,
Elsevier Science. J. Chromatogr.

Table II. Minimum Detection Limits for Certain Organic Compounds by
          HPLC-PID.

Organic Compounds	Minimum Detection Limits (ng)	Lamp Energy(eV)
Bromobenzene	3	10.2
Iodobenzene	4	10.2
Phenol	4	10.2
Chlorobenzene	5	10.2
N,N'-Dimethylaniline	5	9.5
Dioctylphthalate	5	10.2
Fluorobenzene	7	10.2
N,N'-Diethylaniline	20	9.5
N-Methylaniline	60	9.5
o-Xylene	60	10.2
Anthracene	75	9.5
Benzene	150	10.2
Naphthalene	200	9.5
2-Octanone	125	10.2
N-Methylformamide	500	10.2
N,N'-Dimethylformamide	500	10.2
N,N'-Diethylformamide	600	10.2
3-Hexanone	700	10.2

a. HPLC conditions: Alltech C_8 column, 10um, 25 cm x 4.6 mm i.d.;
   mobile phase, acetonitrile-water (75:25, v/v), 0.8 ml/min, directly
   to the photoionization detector in HPLC-PID.

Source: Reproduced with permission from Ref. 116. Copyright 1984,
Elsevier Scientific, Holland.

Post-Column, On-Line, Continuous Photolytic Derivatizations in HPLC-Electrochemical Detection (EC) and FIA-EC

Liquid chromatography (LC) with electrochemical detection (EC) or
LCEC has now been popular for about the past decade, and it has rapidly become a widely accepted and readily applied method of trace organic and inorganic analysis (121-126). Both oxidative and reductive LC-
EC have been described in numerous publications and presentations,
and several firms now offer commercial instrumentation based on both
amperometric and coulometric EC detection for HPLC. As with any
detection method, only those compounds that have inherent EC properties, that can actually be electrochemically oxidized/reduced, will
prove useful, to varying degrees, in LCEC. Those compounds that do
not have natural, inherent EC properties must then be derivatized,
either off-line or on-line, pre- or post-column, to form new materials
with EC detection properties vastly different than the starting, underivatized analyte (127-131). There are some serious disadvantages
in the use of off-line, pre-column derivatizations for any type of
detection in HPLC, and it is generally accepted that on-line, post-
column approaches should be used whenever possible and necessary.
Indeed, various commercial post-column reactors are now on the market
for performing automated, on-line, real-time, continuous chemical

derivatizations in HPLC, especially with UV or FL detection methods
(132). Most of the already described derivatizations for LCEC; however,
have utilized off-line approaches, generally pre-column, and there
have been serious disadvantages in all such approaches/schemes.

To try and overcome this apparent deficiency in derivatizations
for LCEC, and realizing that post-column derivatization methods are
difficult in LCEC using chemical approaches, we have now developed a
quite useful and practical photolysis or photohydrolysis approach in
LCEC (133-136). Figure 6 illustrates the basic HPLC-hv-EC apparatus in
operation, wherein the analytes eluting from the HPLC column are
individually passed around a medium pressure, broad spectrum mercury
discharge lamp immersed in ice-water or a low temperature water bath.
Species that have photolytic or photohydrolytic properties are rapidly
and often very efficiently converted into new products, perhaps stable
inorganic or organic anions, such as nitrite, benzoate, etc., and
these are then passed into the final EC detector. Though, in principle,
it should now be possible to use both reductive and oxidative EC
methods, we have generally found it more practical and simpler to use
oxidative techniques, as for nitrite or sulfide or benzoate anions.
Depending on the particular starting analyte, one or more than one
photoproduct may be formed, and each of these could have different EC
detector properties. Optimization of the system involves determining
the best (longest) residence time for each analyte in the irradiator
Teflon coil or mesh, ideal salt and its concentration in the mobile
phase, mobile phase solvents compatible with hv-EC requirements and
HPLC separations desired for particular analytes, and finally, EC
detector electrode materials, working potentials, single or dual
(parallel or series) arrangements, etc. Once all of the various instrumental requirements are met and compatible, the HPLC-hv-EC system
can be operated routinely for extended periods of time, often with
automated sample injection and data collection. The only limitation
that we have found to date, after several years of operation, has to
do with the fact that the Teflon irradiator tubing tends to harden
with time and crack. This necessitates its replacement, perhaps every
six months, depending on how long and often it has been used. We have
developed a certain woven mesh arrangement for this Teflon irradiator
tubing, which removes most effective dead volume or system variance
due to the post-column photolysis system. This is quite similar in
design to that described earlier by Halasz, Englehardt and Neue, and
others for post-column chemical derivatiztions, on-line in HPLC.
Peak shape is quite similar to any HPLC-UV peak, and overall post-
column variance is little more than without the entire irradiator
in-line between the HPLC column and EC detector entrance (10%).

This newer analytical approach could effectively replace any
need for reductive LCEC in the future, since it appears that most
compounds already studied by reductive means can now be done by
oxidative HPLC-hv-EC methods as well. This has included compounds
such as: 1) organic nitro derivatives, C-nitro, O-nitro, N-nitro, etc.;
2) N-nitroso compounds, such as N-nitrosamines, N-nitroso amino acids,
etc.; 3) organothiophosphates, such as malathion, parathion, ethion,
etc.; 4) aromatic esters and amides, such as benzamide, vitamin B_3,
alkyl benzoates, etc.; 5) beta-lactams, such as penicillins and
cephalosporins; 6) mycotoxins, such as vomitoxin or deoxynivalenol;
8) drugs such as chlordiazepoxide, barbiturates, cocaine, and others.

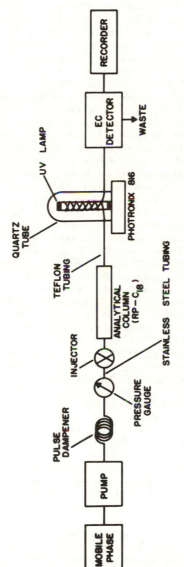

Figure 6. Schematic diagram of the HPLC-hv-EC detection system in operation. Reproduced with permission from Ref. 134. Copyright 1984, American Society for Testing & Materials (ASTM).

We believe that many other classes of compounds will be shown to be suitable analytes for these approaches, and that MDLs will be, as they have been for most of the above compounds, in the low ppb ranges (1-50 ppb or below). In those instances where particular analytes are already amenable to oxidative LCEC, without post-column photolysis, confirmative evidence for their presence can be obtained by HPLC-hv-EC. That is, the dual electrode response ratios for these compounds with and without irradiation are almost always very different, depending on the particular starting analyte. Also, the optimum working electrode potentials with and without irradiation post-column are also very different, providing further confirmation possibilities. We have demonstrated for many compounds additional confirmation via the dual electrode response ratios possible with and without irradiation, using the same electrode working potentials or different sets of these.

    Though not yet commercially available, many suppliers offer irradiation units suitable for this type of post-column photolysis, which only requires the preparation of the Teflon mesh, but even that is already commercially available (Kratos Instruments, Ramsey, N.J.). Various publications have described how to weave one's own Teflon irradiator, and we have summarized such approaches in our own recent publications (133-136). A separate paper will shortly be published dealing with precisely how to weave this type of post-column irradiator, and that should provide sufficient details for anyone to reproduce what we have already done in HPLC-hv-EC (137). In the interim, the excellent work of Engelhardt and Neue, as well as that of Kratos Instruments, should be consulted for further details (138). Any commercially available HPLC system and EC detector can, in principle, be used in HPLC-hv-EC with this approach, and the entire system can be constructed within one single day, if desired. It is easy to learn, easy to construct, easy to operate, and easy to apply to new samples and analytical problems. At the moment, we are investigating a wide variety of licit and illicit drugs, including many barbiturates, and other drugs of abuse. These results are most encouraging and exciting (137). Similarly, we have evidence-at-hand that many organohalogen compounds, including organobromines, organoiodines, and possibly organochlorines, will all be suitable substrates for HPLC-hv-EC methods (139). That could provide the HPLC analyst, at last, with true halogen selective detection methods, especially via the use of dual electrode response ratios and different working electrode surface materials. This work is only now in progress, but already we have sufficient evidence to indicate that all organoiodine and organobromine compounds should be suitable substrates in HPLC-hv-EC. It is our hope and intention that these newer methods of trace organic and inorganic analysis will be adopted by many other analysts and labs, and that they will become standard analytical approaches in the very near future.

The Photoconductivity Detector in HPLC (HPLC-PCD)

This is a detection approach that has been described in the literature for several years, is already commercially available from Tracor Instruments Corp. (Austin, Texas), and has been applied to various classes of compounds (140-143). The detection of non-ionic compounds by electrolytic conductivity depends on the conversion of the original compound to ionic species which, when dissolved in an aqueous electro-

lyte, increases the conductivity of that final electrolyte. It has
long been known that halogenated compounds, organohalogens, such as
many of the commonly used pesticides, are photolytically degraded to
form halogen acids and other products. The photoconductivity detector
(PCD) utilizes the increased conductance of the mobile phase electrolyte which results from the formation of halogen acids when halogenated
compounds are exposed to UV light. Not only halogenated compounds, but
also compounds such as N-nitrosamines, N-nitro derivatives, organothiophosphates (malathion, parathion, etc.), sulfonamides, and a wide
variety of drugs, are all suitable analytes in HPLC-PCD methods.
Figure 7 illustrates a block diagram of the various parts that make-up
a commercial photoconductivity detector in HPLC-PCD operation (142).
In many ways, and perhaps this is why we have included this novel and
selective detector in this review, the PCD resembles the above discussed HPLC-hv-EC approach. The HPLC-hv-EC or photoelectroanalyzer
detection scheme utilizes the exact same photolysis step post-column,
continuous, real-time, on-line, and then passes the newly generated
anions or other photoproducts to the EC detector for measurements.
The PCD does the same thing, but it uses a much more general, non-
selective conductivity measurement approach, which perhaps provides
much less analyte selectivity and information than the photoelectroanalyzer approach. Nevertheless, the PCD has found relative acceptance
and utility, although the number of papers describing its applications
and usefulness are somewhat limited in number. The HPLC mobile phase
requirements for the PCD are probably very similar to those for the
HPLC-hv-EC method, in that mobile phase constituents should not cause
excessively high background noise levels on the conductivity detector.
These same constituents must also not undergo any photolytic process
that could convert them to newer materials then having conductance
properties above the original mobile phase constituents or analytes.
Virtually any HPLC system is compatible with the PCD, though as with
any detector, pressure fluctuations should always be minimized as
much as practical in order to reduce overall noise levels.
      As with the photoelectroanalyzer, the PCD is not compatible with
most normal phase solvents or separations, and thus both methods must
be used with aqueous based mobile phases in reversed phase HPLC. Both
are on-line, real-time, and continuous processes, and both can be used
for long periods of time with high overall reproducibility, accuracy,
and precision. It has always appeared strange that the PCD has never
become as popular as it could or should be, and we believe that this
has been due to a lack of promotion, publications, presentations, and
visibly good applications of general and widespread interest. Basically, it has appeared that the manufacturer has not promoted its own
product, and though they have often presented papers at scientific
meetings, notably the various Pittsburgh Conferences, formal publications in a large number of different journals have been somewhat lack-
ing. The PCD has always been a potentially useful and very practical
detector in HPLC, but it has been treated, or so it would seem, like
an orphan of the manufacturer. Hopefully, this unfortunate state-of-
affairs will soon change, or else the PCD will fade from the market-
place and disappear altogether.

A New Photoelectrochemical Detector for HPLC and FIA (HPLC-PED)

The earlier work with the photoelectroanalyzer system, HPLC-hv-EC, led

us quite naturally to consider placing the light directly within the
EC cell, and onto the working electrode surface (144, 145). The idea
of using electrochemistry to measure or monitor photochemical products,
intermediates, or processes, has been suggested and attempts described
for several years (146-156). Most modern texts on electroanalytical
chemistry have at least a small section dealing with photoelectro-
chemistry, though there are many variations on this term. What we mean
here is the use of photochemistry or light to generate new species
having different electrochemical properties for detection than the
original, unphotolyzed compound/analyte. Though others have used this
approach in solution, using rotating ring-disk electrodes and other
approaches, it is now felt that they were looking at ground-state
intermediates or photoproducts of the photolysis reaction. It seems
unlikely in view of their experimental systems that they were ever
studying the electrochemical properties of electronically promoted
excited states, such as singlets or triplets. Though that earlier work
may still stand on its own, it seems to bear little relation or resem-
blance to that which we are now doing in HPLC-PED or FIA-PED areas.
We should point out, in all fairness and scientific objectivity, that
numerous publications attest to the impossibility of electrochemically
detecting, by conventional oxidative or reductive techniques, any
short-lived, electronically promoted excited states generated photochemically, such as singlets or triplets. Our approach, Figure 8,
utilizes a high intensity, broad spectrum, mercury or xenon/mercury
discharge lamp that produces a light beam of high intensity that is
then focused and directed through the top-half of a two part, thin-
layer, flow-through amperometric transducer cell sold commercially
for more conventional LCEC applications and studies (Bioanalytical
Systems, Inc., West Lafayette, Indiana). The light beam then impinges
through a quartz window cemented into the top half (auxiliary
electrode) of this two part cell, directly onto the working electrode
surface operated at any desired oxidative or reductive potential.
Single or dual, series or parallel arrangements of the cell can be
used, with various working potentials and materials, such as glassy
carbon, gold/mercury, platinu, or silver. In the dual electrode
configuration, these could be of the same or different materials,
one or both could be irradiated, they could be operated at the same
or different working potentials. All of these conditions, as well as
changing the wavelength of light used, increase the overall possible
selectivities available for any given analyte amenable to PED detection (145). Mobile phase requirements are very similar to those
needed for successful LCEC, salts and solvents that have no EC
properties, and no photolytic products derived from these constituents
that would then have their own unique EC properties either. Any
conventional, commercial HPLC system is compatible with the PED, again
requiring relatively pulseless solvent delivery, and the mobile
phase must be thoroughly degassed of dissolved oxygen. This is now
possible via many techniques, but the simplest is surely the use of
a zinc (granular) oxygen scrubber placed on-line, just prior to the
sample injection valve in the HPLC system. All parts of the HPLC-PED
system must then be stainless steel, so that no oxygen from the air
can redissolve in the mobile phase after is has passed the injection
valve stage (157). The light source must be contained in a black box
to protect laboratory workers, and the EC cell should be contained in
a small Faraday cage to reduce background noise levels and spikes.

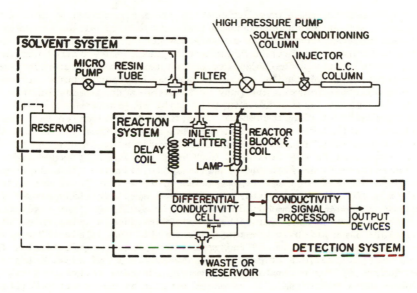

Figure 7. Block diagram showing the relationship of the photo-conductivity detector (PCD) to the other components of the HPLC system. Reproduced with permission from Ref. 142. Copyright 1979, Preston Publications, Inc.

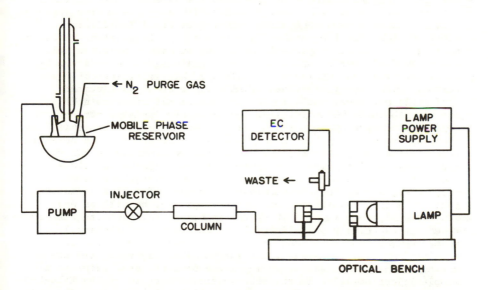

Figure 8. The overall HPLC-photoelectrochemical or FIA-PED detection system using the modified BAS flow-through detection cell. Reproduced with permission from Ref. 144. Copyright 1984, BAS Press, Inc.

This use of a thin-layer, flow-through, amperometric EC cell to generate, in situ, new electrochemically active species appears very different from the work of Weber and colleagues (148, 149). That work used a light beam that passed right across the surface of the working electrode in a wall-jet configuration, but never impinged directly onto the surface at 90° angles. Our own approach puts the light directly onto the working electrode surface at 90°, there is no question that the surface is undergoing irradiation at all times. As the individual HPLC analytes elute from the HPLC column into the PED cell, they are irradiated, presumably at the working electrode surface or just above this, and something is produced that then has its own unique EC properties. None of the compounds in Table III, which is but a partial list of those compounds already shown responsive under PED conditions, have their own EC responses with the lamp off, except for 2-cyclohexen-1-one, a somewhat special case. None of the known solution photo-products of the other compounds in Table III having PED responses with the lamp on, have been found to have any EC responses under these particular operating conditions and potentials. Thus, it would appear that something formed between the ground state starting material and ground state product(s) is responsible for the PED responses being seen. That could be a singlet excited state, a triplet excited state, a ground-state free radical derived from the singlet or triplet after a hydrogen abstraction reaction with the hydrogen donating alcohol solvent, or a radical anion ground-state intermediate leading to the observed final photoreduced products of these ketones and aldehydes. Numerous experiments are still needed to conclusively prove the actual mechanism operative in PED operation. We are acutely aware, indeed painfully aware after certain recent scientific meetings and presentations, that on both experimental and theoretical grounds, it should be impossible to electrochemically detect a short-lived, electronically excited (promoted) singlet or triplet species. Virtually all carbonyl ketones and aldehydes exhibit varying responses on the PED, and MDLs for many of these compounds are equal to or better than the best HPLC-UV conditions. Many compounds that are very poor UV responders appear to have quite usable and practical MDLs via PED.

The PED therefore appears to be a quite sensitive (low ppb) and selective detector that could prove quite useful for many carbonyl derivatives, especially ketones and aldehydes. Together with dual electrode approaches, as suggested above, analyte selectivity appears unique for the compounds amenable to PED. A number of standard chromatograms have already been obtained for mixtures of carbonyls, and actual applications with practical samples are now being developed (158). It is expected that a number of publications in the coming years will more fully describe and depict the PED, as well as what it can be applied for and what unique approaches it may offer the trace analyst. All of this is in addition, of course, to the rather exciting idea of perhaps using electrochemistry to characterize and describe and better understand photochemically promoted, electronically excited states derived from promotion of a ground-state electron into a non-bonding orbital (n-Pi*). Even if it turns out that electrochemistry cannot be used to detect or characterize electronically excited, short-lived species, the PED will still have its own interest as far as what it can be used for in HPLC/FIA applications in trace analytical chemistry.

Table III. Some Organic Compound Responses in FIA-PED[a]

Organic Compound	EC Response (-0.6V)	
	Lamp Off	Lamp On
HPLC mobile phase	no	no
Acetone	no	yes
2-Cyclohexen-1-one	yes	yes
Acetophenone	no	yes
2,3-Butanedione	no	yes
Cinnamaldehyde	no	yes
Mesityl oxide	no	no
Benzaldehyde	no	yes
Indanone	no	yes
o-Aminobenzaldehyde	no	yes
1,4-Cyclohexanedione	no	yes
Fructose	no	no
L-Alanyl-L-phenylalanine	no	no

a. FIA-PED conditions used a mobile phase of 20%MeOH/80% HOH + 0.1M
   NaCl, flow rate of 0.8 ml/min, injection volumes 20ul, -0.6V
   (reductive) vs Ag/AgCl reference electrode, GC working electrode,
   photochemical lamp 500W Hg arc, 10mV full scale EC attenuation.

Conclusions and Future Advances in Selective and Sensitive Detectors
for HPLC and FIA
_____

We have tried to depict some of the more recent advances in the development of new and potentially novel sensitive and selective detectors
of use in HPLC and/or FIA. Our efforts and descriptions have been,
in the main, limited to those areas of most familiarity and experience
in our own research efforts. The development of new approaches to
detection of organics and inorganics in chromatography or flow injection must be guided by certain inherent, basic principles of selection. These could be outlined as follows:
1. The detector should meet an existing need in the analytical
   community, and provide detection limits, methods, and results
   having serious advantages over existing approaches.
2. The detection approach should be widely applicable and usable
   by a very large number of analysts, and it should find ready
   acceptance and utility by that community.
3. The detector should be easy to construct or inexpensive to
   purchase, easy to operate, easy to maintain, easy to learn,
   have a short warm-up time, and be modular in design so that it
   will fit into any laboratory already having HPLC instruments
4. The detection method should be both sensitive and selective,
   in order to find the widest application to real samples, and
   to provide a high degree of analyte identification and/or
   confirmation.
5. The detection method should be quite different from all other
   existing approaches, that is, it should be truly novel and
   provide us with further understanding and knowledge in
   analytical chemistry.
6. The detector should be compatible with both normal and reversed phase chromatography, nonaqueous and aqueous based solvents.
   To some extent, we have chosen to explore new and potentially
novel detection areas, because they seemed to offer promise of unusual
success, and at times, because they seemed to offer solutions to
existing problems of pressing need and concern. Hopefully, these
and other criteria will prevail in the future as well.
   What about the future? We have tried to indicate already some of
the areas where we and others are continuing to develop detection
approaches for HPLC and FIA, and how those developments might proceed
in specific regards. What else is there to be done in the future,
aside from the use of lasers in optical detectors, which some of our
colleagues are already developing or have successfully developed?
Along the lines of the photoelectrochemical detector, we are toying
with the possible uses of electrogenerated luminescence or HPLC/FIA,
as a possible alternative to the now well-established chemiluminescence
post-column detection methods described in recent years (158-164).
Electrogenerated chemiluminescence is very similar to photoelectrochemical detection, in that we are using light and electrochemistry,
but in reversed modes (146, 147). Whereas the PED uses light to generate a new species that then has novel electrochemical properties for
final detection, the ELD approach will use electrochemistry to generate new species that react with one another to generate luminescence,
thereby decaying to neutral species. That luminescence could then be
monitored by any suitable photomultiplier or monochromator. This
approach would clearly be limited to those combinations of compounds

that are able to undergo electrochemical promotion or reduction and
oxidation to produce two oppositely charge species in nearby proximity
at the light output region of the flow-through cell. Whatever light
were produced would then have to be collected, focused, collimated,
and directed into the final light detector. It is possible that indirect or quenched electrogenerated luminescence, a la room temperature phosphorescence in HPLC, could also be devised in the future.
Though certainly a new detection approach in HPLC and FIA, having
inherent novelty, it is questionable at this time if it will have
widespread applicability and acceptance for practical applications.
However, in view of the current decided popularity and application
of chemiluminescence detection in HPLC/FIA, the ELD method just might
rival this now established approach, at least for certain analytes.

Acknowledgments

We wish to thank the editor, S. Ahuja, for the original invitation to
contribute to this ACS Symposium Series publication, and to be able
to enjoy describing some of our recent efforts in the areas of selective and sensitive detection for HPLC/FIA. Much of our own work described here was made possible by various grants and/or contracts
to Northeastern University over the past several years. We wish to
acknowledge firms who have made such commitments, including: Instrumentation Laboratory (Allied Analytical Systems, Inc.), HNU Systems,
Inc., Bioanalytical Systems, Inc., Pfizer, Inc., EM Science, Inc.,
The Barnett Fund for Innovative Research at Northeastern University,
and others, for their continued support, instrumentation donations,
advice, collaboration, interest, encouragement, respect, and most of
all, confidence in what we have been together undertaking. This work
would not have been possible, as well, without the input along the
way of many graduate and undergraduate students, as well as several
Visiting Chinese Scientists from the People's Republic of China. We
especially acknowledge the input and contributions of the following:
D.S. Bushee, B. Karcher, W.R. LaCourse, C.M. Selavka, R.J. Nelson,
J. Burton, Wm. Costa, K-H. Xie, X-D. Ding, L-R. Chen, M. Swartz, S.W.
Jordan, and others. We are especially grateful to our colleagues
within The Barnett Institute and Department of Chemistry for encouraging us along the way, making many valued and valuable suggestions,
and for constantly taking an interest in the progress of the R&D
programs described above. Special gratitude and acknowledgment goes
to B.L. Karger, W.C. Giessen, P.L. Vouros, and R.W. Giese, all at NU.
    This is contribution number 244 from The Barnett Institute at NU.

Literature Cited

1.   Scott, R.P.W. "Liquid Chromatography Detectors"; Elsevier
     Scientific Publishing Co.: Amsterdam, Holland, 1977.
2.   Kissinger, P.T.; Felice, L.J.; Miner, D.J.; Preddy, C.R.;
     Shoup, R.E. In "Contemporary Topics in Analytical and Clinical
     Chemistry, Volume 2"; Hercules, D.M.;Hieftje, G.M.; Snyder, L.R.;
         Evenson, M.A., Eds.; Plenum Press, Inc.: New York, 1978;
     Chap. 3.
3.   Snyder, L.R.;   Kirkland, J.J. "Introduction to Modern Liquid
     Chromatography" Second Edition; J. Wiley & Sons, Inc.: New York,
     1979; Chap. 4.

4. Vickrey, T.M., Ed. "Liquid Chromatography Detectors"; Marcel
    Dekker: New York, 1983.
5. Yeung, E.S. In "Advances in Chromatography, Volume 23"; Giddings,
    J.C.; Grushka, E.; Cazes, J.; Brown, P.R., Eds.; Marcel Dekker:
    New York, 1984; Chap. 1.
6. Hancock, W.S.; Sparrow, J.T. "HPLC Analysis of Biological Com-
    pounds, A Laboratory Guide"; Marcel Dekker: New York, 1984;
    Chap. 4.
7. Kissinger, P.T., Ed. "An Introduction to Detectors for Liquid
    Chromatography" First Edition; BAS Press: West Lafayette, Indi-
    ana, 1981.
8. Lawrence, J.F. "Organic Trace Analysis by Liquid Chromatography";
    Academic Press: New York, 1981; Chap. 5.
9. DiCesare, J.L.; Ettre, L.S. J. Chromatogr., Chromatogr. Revs.
    1982, 251, 1.
10. Varadi, M.; Balla, J.; Pungor, E. Pure & Appl. Chem. 1979, 51,
    1177.
11. Borman, S.A. Anal. Chem. 1982, 54, 327A.
12. Brenner, M.; Sims, C.W. American Laboratory (Fairfield, Ct.)
    1981, 13(10), 78.
13. Ettre, L.S. J. Chromatogr. Sci. 1978, 16, 396.
14. Wise, S.A.; May, W.E. Research/Development Magazine 1977, 54
    (October).
15. Krull, I.S.; Lankmayr, E.P. American Laboratory (Fairfield, Ct.)
    1982, 14(5), 18.
16. Xie, K-H.; Colgan, S.; Krull, I.S. J. Liquid Chrom. 1983, 6(S-2),
    125.
17. Colgan, S.; Krull, I.S. In "Post-Column Reaction Detectors in
    HPLC" Krull, I.S., Ed.; Marcel Dekker: New York, 1985, in press.
18. Brinckman, F.E.; Bernhard, M., Eds. "The Importance of Chemical
    Speciation in Environmental Processes"; Springer-Verlag: West
    Berlin; Heidelberg, 1985, in press.
19. Krull, I.S. In "Environmental Analysis by Liquid Chromatography";
    Lawrence, J.F., Ed.; Humana Press: Clifton, N.J., 1984; Chap. 5.
20. Krull, I.S.; Jordan, S. American Laboratory (Fairfield, Ct.)
    1980, 12(10), 21.
21. Krull, I.S. Trends in Anal. Chem. 1984, 3(3), 76.
22. Krull, I.S.; Panaro, K; Gershman, L.L. J. Chromatogr. Sci. 1983,
    21, 460.
23. Berman, E. "Trace Metals and Their Analysis"; Heyden: London,
    1980.
24. Risby, T.H., Ed. "Ultratrace Metal Analysis in Biological Science
    and Environment"; Advances in Chemistry Series 171; American
    Chemical Society: Washington, D.C., 1979.
25. Jewett, K.L.; Brinckman, F.E. In "Detectors in Liquid Chromatog-
    raphy"; Vickrey, T.M., Ed.; Marcel Dekker: New York, 1983; Chap.6.
26. Schwedt, G. "Chromatographic Methods in Inorganic Analysis";
    A.H. Verlag: Heidelberg, 1981.
27. Uden, P.C.; Bigley, I.E.; Walters, F.H. Anal. Chim. Acta 1978,
    100, 555.
28. Bushee, D.S.; Krull, I.S.; Demko, P.R.; Smith, S.B., Jr. J.
    Liquid Chrom. 1984, 7, 861.
29. Cassidy, R.M. In "Trace Analysis, Volume 1"; Lawrence, J.F., Ed.;
    Academic Press: New York, 1981; p. 122.
30. MacCrehan, W.A.; Durst, R.A. Anal. Chem. 1978, 50, 2108.
31. MacCrehan, W.A.; Durst, R.A. Anal. Chem. 1981, 53, 1700.
32. MacCrehan, W.A. Anal. Chem. 1981, 53, 74.

33. Kissinger, P.T.; Heineman, W.R., Eds. "Laboratory Techniques in Electroanalytical Chemistry"; Marcel Dekker: New York, 1984.
34. Shoup, R.E., Ed. "Recent Reports on Liquid Chromatography/ Electrochemistry"; BAS Press: West Lafayette, Indiana, 1982.
35. Messman, J.D.; Rains, T.C. Anal. Chem. 1981, 53, 1632.
36. Chakraborti, D.; Hillman, D.C.J.; Irgolic, K.J.; Zingaro, R.A. J. Chromatogr. 1982, 249, 81.
37. Kahn, N.; Van Loon, J.C. J. Liquid Chrom. 1979, 2, 23.
38. Van Loon, J.C. Anal. Chem. 1979, 51, 1139A.
39. Van Loon, J.C. American Laboratory (Fairfield, Ct.) 1981, 13(5), 47.
40. Renoe, B.W.; Shideler, C.E.; Savory, J. Clin. Chem. 1981, 27, 1546.
41. Tittarelli, P.; Mascherpa, A. Anal. Chem. 1981, 53, 1466.
42. Koizumi, H.; McLaughlin, R.D.; Hadeishi, T. Anal. Chem. 1979, 51, 387.
43. Fernandez, F.J. Perkin-Elmer Atomic Absorption Newsletter 1977, 16(2), 33.
44. Koropchak, J.A.; Coleman, G.N. Anal. Chem. 1980, 52, 1252.
45. Roden, D.R.; Tallman, D.E. Anal. Chem. 1982, 54, 307.
46. Mackey, D.J. J. Chromatogr. 1982, 237, 79.
47. Fish, R.H.; Komlenic, J.J. Anal. Chem. 1984, 56, 510.
48. Slavin, W.; Schmidt, G.J. J. Chromatogr. Sci. 1979, 17, 610.
49. Jewett, K.L.; Brinckman, F.E. J. Chromatogr. Sci. 1981, 19, 583.
50. Brinckman, F.E.; Blair, W.R.; Jewett, K.L.; Iverson, W.P. J. Chromatogr. Sci. 1977, 15, 493.
51. Parks, E.J.; Brinckman, F.E.; Blair, W.R. J. Chromatogr. 1979, 185, 563.
52. Brinckman, F.E.; Jewett, K.L.; Iverson, W.P.; Irgolic, K.J.; Ehrhardt, K.C.; Stockton, R.A. J. Chromatogr. 1980, 191, 31.
53. Mazzo, D.J.; Elliott, W.G.; Uden, P.C.; Barnes, R.M. Appl. Spec. 1984, 38, 585.
54. Krull, I.S.; Panaro, K.W. Appl. Spec. 1985, 39, in press.
55. Jinno, K.; Tsuchida, H.; Nakanishi, S.; Hirata, Y.; Fujimoto, C. Appl. Spec. 1983, 37, 258.
56. Gardner, W.S.; Landrum, P.F.; Yates, D.A. Anal. Chem. 1982, 54, 1196.
57. Jinno, K.; Tsuchida, H. Anal. Lett. 1982, 15, 427.
58. Morita, M.; Uehiro, T. Anal. Chem. 1981, 53, 1997.
59. Gast, C.H.; Kraak, J.C.; Poppe, H.; Maessen, F.J.M.J. J. Chromatogr. 1979, 185, 549.
60. Fraley, D.M.; Yates, D.A.; Manahan, S.E.; Stalling, D.; Petty, J. Appl. Spec. 1981, 35, 525.
61. Fraley, D.M.; Yates, D.; Manahan, S.E. Anal. Chem. 1979, 51, 2225.
62. Morita, M.; Uehiro, T.; Fuwa, K. Anal. Chem. 1980, 52, 349.
63. Whaley, B.S.; Snable, K.R.; Browner, R.F. Anal. Chem. 1982, 54, 162.
64. Heine, D.R.; Denton, M.B.; Schlabach, T.D. Anal. Chem. 1982, 52, 81.
65. Yoshida, K.; Hasegawa, T.; Haraguchi, H. Anal. Chem. 1983, 55, 2106.
66. Van Loon, J.C. Spectrochim. Acta 1983, 38B, 1509.
67. Hausler, D.W.; Taylor, L.T. Anal. Chem. 1981, 53, 1227.
68. Hausler, D.W.; Taylor, L.T. Anal. Chem. 1981, 53, 1223.
69. Irgolic, K.J.; Stockton, R.A.; Chakraborti, D.; Beyer, W. Spectrochim. Acta 1983, 38B, 437.

70. Lawrence, K.E.; Rice, G.W.; Fassel, V.A. Anal. Chem. 1984, 56, 289.
71. Brown, R.J.; Biggs, W.R. Anal. Chem. 1984, 56, 646.
72. Krull, I.S.; Bushee, D.; Savage, R.N.; Schleicher, R.G.; Smith, S.B., Jr. Anal. Lett. 1982, 15(A3), 267.
73. Bushee, D.; Krull, I.S.; Savage, R.N.; Smith, S.B., Jr. J. Liquid Chrom. 1982, 5, 563.
74. White, P.C. Analyst 1984, 109, 677.
75. Ibrahim, M.; Gilbert, T.W.; Caruso, J.A. J. Chromatogr. Sci. 1984, 22, 111.
76. McCarthy, J.P.; Caruso, J.A.; Fricke, F.L. J. Chromatogr. Sci. 1983, 21, 389.
77. Carnahan, J.W.; Mulligan, K.J.; Caruso, J.A. Anal. Chim. Acta 1981, 130, 227.
78. Ng, K.C.; Caruso, J.A. Anal. Chem. 1983, 55, 2032.
79. Kirkbright, G.F.; Li-Xing, Z. Analyst, 1982, 107, 617.
80. Browner, R.F.; Boorn, A.W. Anal. Chem. 1984, 56, 875A.
81. Browner, R.F.; Boorn, A.W. Anal. Chem. 1984, 56, 786A.
82. Nixon, D.E.; Fassel, V.A.; Kniseley, R.N. Anal. Chem. 1974, 46, 210.
83. Swaidan, H.M.; Christian, G.D. Anal. Chem. 1984, 56, 120.
84. Azia, A.; Broekaert, J.A.C.; Leis, F. Spectrochim. Acta 1982, 37B, 369.
85. Ebdon, L.; Wilkinson, J.R.; Jackson, K.W. Anal. Chim. Acta 1982, 136, 191.
86. Ricci, G.R.; Shepard, L.S.; Colovos, G.; Hester, N.E. Anal. Chem. 1981, 53, 610.
87. Ebdon, L. Proc. 9th Fed. Anal. Chem. Spec. Soc. Mtg., 1982, no. 47.
88. Bushee, D.S.; Krull, I.S.; Schleicher, R.G.; Smith, S.B., Jr. Proc. 1985 Pitts. Conf. Anal. Chem. Appl. Spec. Mtg., 1985, no. 638.
89. Prack, E.R.; Bastiaans, G.J. Anal. Chem. 1983, 55, 1654.
90. Black, M.S.; Thomas, M.B.; Browner, R.F. Anal. Chem. 1981, 53, 2224.
91. Black, M.S.; Browner, R.F. Anal. Chem. 1981, 53, 249.
92. Skogerboe, R.K.; Dick, D.L.; Pavlica, D.A.; Lichte, F.E. Anal. Chem. 1975, 47, 568.
93. Summerhays, K.D.; Lamothe, P.J.; Fries, T.L. Appl. Spec. 1983, 37, 25.
94. Karcher, B.; Krull, I.S.; Schleicher, R.G.; Smith, S.B., Jr., unpublished data.
95. Boorn, A.W.; Browner, R.F. Anal. Chem. 1982, 54, 1402.
96. Barrett, P.; Pruszkowska, E. Anal. Chem. 1984, 56, 1927.
97. Novak, J.W., Jr.; Browner, R.F. Anal. Chem. 1980, 52, 792.
98. Goulden, P.D.; Anthony, D.H.J. Anal. Chem. 1982, 54, 1678.
99. Willoughby, R.C.; Browner, R.F. Anal. Chem. 1984, 56, 2626, 2702.
100. Farino, J.; Browner, R.F. Anal. Chem. 1984, 56, 2709.
101. Voyksner, R.D.; Bursey, J.T.; Pellizzari, E.D. Anal. Chem. 1984, 56, 1507.
102. Covey, T.; Henion, J. Anal. Chem. 1983, 55, 2275.
103. Hardin, E.D.; Fan, T.P.; Blakley, C.R.; Vestal, C.L. Anal. Chem. 1984, 56, 2.
104. Pilosof, D.; Kim, H.Y.; Dyckes, D.F.; Vestal, M.L. Anal. Chem. 1984, 56, 1236.
105. Vouros, P.; Lankmayr, E.P.; Hayes, M.; Karger, B.L.; McGuire, J.M. J. Chromatogr. 1982, 251, 175.

106. Hayes, M.J.; Lankmayr, E.P.; Vouros, P.; Karger, B.L.; McGuire, J.M. Anal. Chem. 1983, 55, 1745.
107. Karcher, B.; Krull, I.S.; Schleicher, R.G.; Smith, S.B., Jr., unpublished data.
108. Krull, I.S.; Swartz, M.E.; Driscoll, J.N. In "Advances in Chromatography, Volume 24"; Giddings, J.C.; Grushka, E.; Cazes, J.; Brown, P.R., Eds.; Marcel Dekker: New York, 1984; Chap. 8.
109. Langhorst, M.L.; Nestrick, T.J. Anal. Chem. 1979, 51, 2018.
110. Langhorst, M.L. J. Chromatogr. Sci. 1981, 19, 98.
111. Kapila, S.; Bornhop, D.J.; Manahan, S.E.; Nickell, G.L. J. Chromatogr. 1983, 259, 205.
112. Driscoll, J.N. J. Chromatogr. Sci. 1982, 20, 91.
113. Jaramillo, L.F.; Driscoll, J.N. J. Chromatogr. 1979, 186, 637.
114. Krull, I.S.; Swartz, M.; Hillard, R.; Xie, K-H.; Driscoll, J.N. J. Chromatogr. 193, 260, 347.
115. Towns, B.D.; Driscoll, J.N. American Laboratory (Fairfield, Ct.) 1982, 14(7), 56.
116. Driscoll, J.N.; Conron, D.W.; Ferioli, P.; Krull, I.S.; Xie, K-H. J. Chromatogr. 1984, 302, 43.
117. Driscoll, J.N.; Becker, H. Proc. 1977 Pitts. Conf. Anal. Chem. Appl. Spec. Mtg., no. 387.
118. Locke, D.C.; Dhingra, B.S.; Baker, A.D. Anal. Chem. 1982, 54, 447.
119. Schmermund, J.T.; Locke, D.C. Anal. Lett. 1975, 8(9), 611.
120. Krull, I.S.; Nelson, R.; Driscoll, J.N., unpublished data.
121. Krull, I.S.; Bratin, K.; Shoup, R.E.; Kissinger, P.T.; Blank, C.L. American Laboratory (Fairfield, Ct.) 1983, 15(2), 57.
122. Bratin, K.; Blank, C.L.; Krull, I.S.; Lunte, C.E.; Shoup, R.E. American Laboratory (Fairfield, Ct.), 1984, 16(5), 33.
123. Kissinger, P.T., Ed. "Principles and Applications of Liquid Chromatography/Electrochemistry"; BAS Press: West Lafayette, Indiana, 1984.
124. Kissinger, P.T. In "Laboratory Techniques in Electroanalytical Chemistry"; Kissinger, P.T.; Heineman, W.R., Eds.; Marcel Dekker: New York, 1984; Chap. 22.
125. Krstulovic, A.M.; Colin, H.; Guiochon, G.A. In "Advances in Chromatography, Volume 24"; Giddings, J.C.; Grushka, E.; Cazes, J.; Brown, P.R., Eds.; Marcel Dekker: New York, 1984; Chap. 4.
126. Brunt, K. In "Trace Analysis, Volume 1"; Lawrence, J.R., Ed.; Academic Press: New York, 1981, p. 47.
127. Kissinger, P.T.; Bratin, K.; Davis, G.C.; Pachla, L.A. J. Chromatogr. Sci. 1979, 17, 137.
128. Jacobs, W.A.; Kissinger, P.T. J. Liquid Chrom. 1982, 5(4), 669.
129. Jacobs, W.S.; Kissinger, P.T. J. Liquid Chrom. 1982, 5(5), 881.
130. Allison, L.A.; Mayer, G.S.; Shoup, R.E. Anal. Chem. 1984, 56, 1089.
131. Mahachi, T.J.; Carlson, R.M.; Poe, D.P. J. Chromatogr. 1984, 298, 279.
132. Frei, R.W. In "Chemical Derivatization in Analytical Chemistry, Volume 1: Chromatography"; Frei, R.W.; Lawrence, J.F., Eds.; Plenum Press: New York, 1981; Chap. 4.
133. Ding, X-D.; Krull, I.S. J. Agric. Food Chem. 1984, 32, 622.
134. Krull, I.S.; Selavka, C.; Ding, X-D.; Bratin, K.; Forcier, G. J. Forensic Sci. 1984, 29(2), 449.

135. Krull, I.S.; Ding, X-D.; Selavka, C.; Nelson, R. LC Magazine 1984, 2(3), 214.
136. Selavka, C.M.; Nelson, R.J.; Krull, I.S.; Bratin, K. J. Pharm. Biomed. Anal. 1985, 3, in press.
137. Selavka, C.M.; Krull, I.S., unpublished data.
138. Neue, U-D. Ph.D. Thesis, University of Saarbrucken, Federal Republic of Germany, 1976.
139. Colgan, S.; Selavka, C.M.; Burton, J.; Krull, I.S., unpublished data.
140. Jasinski, J.S. Anal. Chem. 1984, 56, 2214.
141. Walters, S.M. J. Chromatogr. 1983, 259, 227.
142. Popovich, D.J.; Dixon, J.B.; Ehrlich, B.J. J. Chromatogr. Sci. 1979, 17, 643.
143. McKinley, W.A. J. Anal. Toxicol. 1981, 5, 209.
144. LaCourse, W.R.; Krull, I.S.; Bratin, K.; Forcier, G. Proc. 5th Intl. Sym. LCEC Voltamm. Mtg., 1984, no. 12.
145. LaCourse, W.R.; Krull, I.S.; Bratin, K. Anal. Chem. 1985, 57, submitted for publication.
146. Bard, A.J.; Faulkner, L.R. "Electrochemical Methods: Fundamentals and Applications"; J. Wiley: New York, 1980; Chap. 14.
147. Tachikawa, H.; Faulkner, L.R. In "Laboratory Techniques in Electroanalytical Chemistry"; Kissinger, P.T.; Heineman, W.R., Eds.; Marcel Dekker: New York, 1984; Chap. 23.
148. Weber, S.G.; Morgan, D.M.; Elbicki, J.M. Clin. Chem. 1983, 29(9), 1665.
149. Weber, S.G. U.S. Patent   4 293 310, October 6, 1981.
150. Lubbers, J.R.; Resnick, E.W.; Gaines, P.R.; Johnson, D.C. Anal. Chem. 1974, 46, 865.
151. Johnson, D.C.; Resnick, E.W. Anal. Chem. 1972, 44, 637.
152. Kuwana, T. In "Electroanalytical Chemistry, A Series of Advances"; Bard, A.J., Ed.; Volume 1; Marcel Dekker: New York, 1966; p. 197.
153. Finklea, H.O. J. Chem. Educ. 1983, 60(4), 325.
154. Perone, S.P.; Birk, J.R. Anal. Chem. 1966, 38, 1589.
155. Patterson, J.I.H.; Perone, S.P. J. Phys. Chem. 1973, 77, 2427.
156. Jamisson, R.A.; Perone, S.P. J. Phys. Chem. 1972, 76, 330.
157. MacCrehan, W.A.; May, W.E. Anal. Chem. 1984, 56, 625.
158. Nelson, J.K.; Getty, R.H.; Birks, J.W. Anal. Chem. 1983, 55, 1767.
159. Sigvardson, K.W.; Kennish, J.M.; Birks, J.W. Anal. Chem. 1984, 56, 1096.
160. Weinberger, R.; Mannan, C.A.; Cerchio, M.; Grayeski, M.L. J. Chromatogr. 1984, 288, 445.
161. Sigvardson, K.W.; Birks, J.W. Anal. Chem. 1983, 55, 432.
162. Weinberger, R. J. Chromatogr. 1985, in press.
163. Koziol, T.; Grayeski, M.L.; Weinberger, R. J. Chromatogr. 1985, in press.
164. Weinberger, R.; Koziol, T.; Millington, G. Chromatographia 1985, in press.

RECEIVED October 8, 1985

Microcomputer-Assisted Retention Prediction in Reversed-Phase Liquid Chromatography

Kiyokatsu Jinno and Kazuya Kawasaki

School of Materials Science, Toyohashi University of Technology, Toyohashi 440, Japan

A computer-assisted system for predicting retention of
aromatic compounds has been investigated in reversed-
phase liquid chromatography. The basic retention descriptions have been derived from the studies on quantitative structure-retention relationships. The system
was constructed on a 16-bit microcomputer and then
evaluated by comparing the retention data between
measured and predicted values. The excellent agreement between both values were observed on an octadecylsilica stationary phase with acetonitrile and methanol
aqueous mobile phase systems. This system has been
modified to give us the information for optimal separation conditions in reversed-phase separation mode. The
approach could also work well for any other reversed-
phase stationary phases such as octyl, phenyl and ethyl
silicas.

Although liquid chromatography was discovered about 80 years ago, it
has been widely used as an efficient analytical techniques only during
the last decade. With the development of high performance liquid
chromatography (LC), complex mixtures can now be readily separated
into their components and then analyzed quantitatively and qualitatively. The rapid ascent of this technique is one of the most impressive phenomena in the history of analytical chemistry. Furthermore,
the utility of LC has been catalyzed by the development of modern high
efficiency columns, sensitive and specific on-line detection systems,
and microprocessors for instrument operations, as well as data
handling. However, the selection of satisfactory separation conditions is still a major problem in LC. In order to set the best separation condition, analysts generally survey the information about it
for their respective purposes in many published articles, and then
find the most promising combination of mobile phase and stationary
phase. After that, trial and error experiments are performed to find
the best optimal separation condition for their purposes. This
approach is very difficult and time-consuming. An improved approach
involves systematic optimization with the computer techniques such as

0097-6156/86/0297-0167$06.25/0

"Window Diagram" or "Simplex" methods (1-5). This approach is very useful to get optimal separation conditions; however, it has one major limitation. It necessitates all materials of interest be available. In practice, it is difficult to achieve this because standard materials may not be commercially available or are highly toxic and therefore can create new pollution problems. To overcome this disadvantage, an alternate approach has been proposed by the authors (6-10). That attempt is retention prediction. If retention of solutes can be predicted at appropriate experimental condition, optimization procedure can be more easily attained in a short time.

In order to predict retention of any solute, a clear understanding of the retention mechanism is needed. During recent years, much effort has been directed to investigate the mechanism of solute retention in reversed-phase LC both by chromatography and/or spectroscopy (11-18). At present, "solvophobic theory" introduced by Horváth et al. (19-21) is generally acknowledged as one of the most consistent theories to describe solute distribution phenomena in reversed-phase LC. According to this entropically driven interaction model, it can be anticipated that physicochemical parameters such as solute's surface area, its partition coefficient between two immiscible phases and its aqueous solubility may correlate with the retention in reversed-phase LC. In practice, such correlations between retention and physicochemical parameters exist and some of these parameters have been determined based on those relationships. For example, a great deal of effort have been made to use reversed-phase LC as the most promising method for the determination of logarithm of partition coefficient of a compound in 1-octanol/water system which is a good measure of its hydrophobicity: this is an important parameter in quantitative structure-activity relationships (QSAR) (22-28). This idea in QSAR can be applied to predict retention in reversed-phase LC.

That is to say, prediction of retention in reversed-phase LC can be made, based on the premise that relationships exist between the physicochemical parameters representing the molecular properties of the solute such as structure, shape and/or electronic states etc., and its retention, if such parameters are available. The basic concept is shown in equation-1:

$$k' = f\ (P_i) \tag{1}$$

where k' is a capacity factor of a solute defined as $k'=(t_r-t_o)/t_o$, in which t_r is the retention time of it and t_o is the column void-time and P_i is physicochemical parameters of the solute. Such approach has been named as quantitative structure-retention relationships (QSRR) by Horváth et al. (29). Based on the QSRR studies, one will be able to derive equations such as equation-1 in various reversed-phase LC systems. And then, by using the equation obtained, retention of solutes can be predicted.

In this contribution, we will describe the basic approach to construct the retention prediction system in reversed-phase LC for alkylbenzenes, polycyclic aromatic hydrocarbons (PAHs) and polar group substituted benzenes, based on the use of such established relationships between retention and physicochemical parameters of these compounds. The system has been constructed on a 16-bit microcomputer, and the application for optimization of separation conditions will be demonstrated.

Experimental

The liquid chromatographic system used consisted of a Microfeeder MF-2 (Azuma Electric, Co., Ltd., Tokyo, Japan) as the pump and a Uvidec-100III ultraviolet spectrometer (Jasco, Tokyo, Japan) as the detector set at 210 nm and/or 260 nm. The columns were placed in a Komatsu DW 620 (Tokyo, Japan) thermostat kept at 20 $\pm$ 0.1°C.

Four reversed-phase columns (PTFE tubings of 0.5 mm i.d. x 12 cm long) were prepared by the slurry technique; (1) Jasco FineSIL C18 (5 μm), (2) Develosil C8 (5 μm, Nomura Chemicals, Seto, Japan), (3) Develosil C-phenyl (10 μm, Cp) and (4) Jasco FineSIL C2 (10 μm).

The mobile phases consisted of HPLC grade acetonitrile and methanol purchased from Kanto Chemicals (Tokyo, Japan) and purified water. All of test substances were commercially available from many sources. The flow rate of the mobile phase was always 4 μL/min. Prior to the measurements the columns were equilibrated with each mobile phase until a constant value was obtained for the retention of the test substance. The test solutes were analyzed as solutions in each mobile phase, in a concentration of a few hundred parts per million (ppm). If a solute was not easily soluble in the mobile phase, it was dissolved in pure acetonitrile or methanol depending on the mobile phase in use.

There have been a number of publications on determination of column void-volume (30-35) and this subject is still one of the hot topics in LC. From our knowledge and basic study on this subject (36, 37), we decided sodium nitrite and sodium nitrate are good solutes to measure column void-time in reversed-phase LC. Therefore, the elution time of sodium nitrite was used as t_0 in this work. Its concentration used was c.a. 100 ppm in each mobile phase.

All measurements of capacity factors for aromatic compounds with various LC systems were made at least in triplicates. The average reproducibility of each run was better than 1.0 % relative standard deviation.

Multiple regression analyses were performed by the use of a MELCOM 800 computer (Mitsubishi Electric, Co., Ltd., Osaka, Japan). The computer system for retention prediction was a 16-bit microcomputer NEC 9801 (Nippon Electric, Co., Ltd., Tokyo, Japan), and the programs were written in BASIC language.

Results and Discussion

In the present study, the capacity factor values of 20 alkylbenzenes, 18 PAHs and 28 polar groups substituted benzenes were determined on the four reversed-phase columns with mobile phases of methanol-water and acetonitrile-water.

In order to establish the relationships between retention and various physicochemical parameters, here we call "descriptors", as describing the features of molecular properties of a solute, the followings were selected:

log P ; it is the logarithm of the partition coefficient in 1-octanol/water and the measure of the hydrophobicity of a molecule (38).

π ; it is the hydrophobic substituent constant as the measure of the hydrophobicity of the substituted group in a molecule (39).

V_w, A_w; they are calculated from the van der Waals radii of the atoms composing a molecule (40). V_w is the van der Waals volume and

A_w is the van der Waals surface area.

$\underline{F}_c$; correlation factor, F_c is calculated as $F_c=$(number of double bonds) + (number of primary and secondary carbons) - 0.5 for non-aromatic ring (41,42).

$\underline{\sigma}_h$; Hammett's constant.

$\underline{HA},\underline{HD}$; they are parameters showing the extent of substituted groups; HA is equal to the number of hydrogen acceptor groups and HD is equal to the number of hydrogen donar groups in a molecule (39).

From the basic studies on QSRR for alkylbenzenes, PAHs and polar groups substituted benzenes, the descriptors stating as follows are highly promising to describe the retention in reversed-phase LC with the four different stationary phases: for alkylbenzenes and PAHs, F_c and log P are good descriptors for describing their retention with C18 and Cp columns, and A_w and log P with C8 and C2 columns, respectively. For substituted benzenes except phenols, π and (HA-HD) are good descriptors. In this instance, (HA-HD) indicates the hydrogen accepting ability of the whole molecule. For phenols, π and σ_h are good descriptors with the four columns. By the above results, the equations listed in Table I were obtained by the multiple regression analyses for retention description of aromatic compounds. The equations are only the retention descriptions at the fixed mobile phase compositions, however.

To generalize the retention descriptions for n different mobile phase compositions, the following n equations should be obtained by the same procedures with multiple regression analysis. The relationship for each compound's group and system can be described in the general form as follows:

$X=X_1$ $\log k' = a_1 P_1 + b_1 P_2 + c_1$

$X=X_2$ $\log k' = a_2 P_1 + b_2 P_2 + c_2$

$\vdots$ $\vdots$

$X=X_n$ $\log k' = a_n P_1 + b_n P_2 + c_n$ (2')

where X is the volume fraction of organic modifier in the mobile phase and a and b are the coefficients corresponding descriptors, P_1 and P_2, respectively and c is the intercept, and n is the number of examined experimental conditions.

If a, b and c can be expressed as functions of X, namely, if X-a, X-b and X-c are highly correlated, the following three equations can be obtained by the multiple regression analyses:

$$a = f_1 (X) = \sum_{i=1}^{m} d_i X^i \qquad (3)$$

$$b = f_2 (X) = \sum_{i=1}^{m} e_i X^i \qquad (4)$$

$$c = f_3 (X) = \sum_{i=1}^{m} f_i X^i \qquad (5)$$

where m is the number less than n and d_i, e_i and f_i are the coefficients corresponding to "i" powers of X. If such equations as 3, 4 and 5 can be derived, the following equation-6 will be obtained for each compounds' group.

$$\log k' = f_1(X)P_1 + f_2(X)P_2 + f_3(X) \qquad (6)$$

Table I. Retention descriptions by multi-combination of two descrip-
          tors for aromatic compounds, where the following relation-
          ship is assumed:

$$\log k' = a\, P_1 + b\, P_2 + c \qquad\qquad (2)$$

(a) alkylbenzenes and PAHs

column and mobile phase*1		P_1	P_2	a	b	c	r*2
C18	A	log P	F_c	0.187	0.029	−0.552	0.990
C18	M	log P	F_c	0.230	0.056	−0.847	0.989
C8	A	A_w	log P	0.070	0.067	−0.331	0.983
C8	M	A_w	log P	0.107	0.095	−0.634	0.989
Cp	A	log P	F_c	0.235	−0.022	−0.213	0.995
Cp	M	log P	F_c	0.325	−0.016	−0.663	0.996
C2	A	A_w	log P	0.049	0.090	−0.284	0.979
C2	M	A_w	log P	0.010	0.056	−0.829	0.982

(b) polar groups substituted benzenes except phenols

column and mobile phase		P_1	P_2	a	b	c	r
C18	A	π	(HA−HD)	0.136	−0.049	−0.007	0.959
C18	M	π	(HA−HD)	0.130	−0.048	−0.050	0.925
C8	A	π	(HA−HD)	0.240	−0.031	0.215	0.965
C8	M	π	(HA−HD)	0.283	−0.045	0.117	0.978
Cp	A	π	(HA−HD)	0.178	−0.025	0.222	0.976
Cp	M	π	(HA−HD)	0.191	0.005	0.106	0.982
C2	A	π	(HA−HD)	0.234	−0.017	0.212	0.950
C2	M	π	(HA−HD)	0.273	−0.010	−0.098	0.982

(c) phenols

column and mobile phase		P_1	P_2	a	b	c	r
C18	A	π	S*3	0.182	0.182	−0.368	0.952
C18	M	π	S	0.166	0.175	−0.343	0.974
C8	A	π	S	0.294	0.080	−0.192	0.952
C8	M	π	S	0.343	0.099	−0.235	0.979
Cp	A	π	S	0.220	0.131	−0.221	0.960
Cp	M	π	S	0.191	0.158	−0.401	0.970
C2	A	π	S	0.302	0.115	−0.103	0.953
C2	M	π	S	0.312	0.152	−0.171	0.956

*1 A: acetonitrile:water = 65 : 35.
   M: methanol:water = 75 : 25.

*2 r: correlation coefficient.

*3 $S = \pi \sigma_h (1-\pi)$.

This equation means that, if X, the concentration of organic modifier in the mobile phase, and P_1 and P_2, descriptors of a compound are given, the logarithm of the capacity factor, log k' can be determined for any chromatographic conditions. This is the basic concept of the retention prediction system investigated in this study.

To simplify the explanation, and as the typical example for constructing the retention prediction system, only the case of the C18 column has been focused in the following discussions.

The equations obtained actually by the experiments for the C18 column are shown in Table IV, where the data listed in Table II and Table III were used to obtain them. Given the equations for the C18 column, it is possible to predict retention of aromatic compounds. Some compounds were probed to measure their retention and the measured capacity factors were compared with the values predicted by using the equations in Table IV. It can be found in Figure 1 that predicted capacity factors for almost all compounds agree with the measured ones within ± 10 % relative errors. Of course, this fairly good agreement between measured and predicted capacity factor values with the C18 column suggests that it should be possible to predict retention on other reversed-phases such as C8, Cp and C2 columns.

According to the above mentioned procedures, the computer-assisted retention prediction system (RPS) for the C18 column was constructed on the 16-bit microcomputer. The flow-chart of this function of RPS is shown in Figure 2. In the use of RPS as the system to predict retention times of solutes, the following data are input with the interactive style after accessing the function on the CRT of the computer; (1) the compound name or the chemical formula of interesting solutes, (2) experimental conditions (mobile phase (M), volume fraction of organic modifier in the mobile phase (X), flow rate of the mobile phase (F)).

When compounds' names or chemical formulas are input to RPS, the computer calculates suitable descriptors for the input compounds' groups and then capacity factors (k') for the solutes at selected mobile phase composition are predicted.

From the input values of F, the theoretical plate number, N is calculated by the relationship-7 approximately modified for the column used (experimentally determined), though generally N is a function of linear velocity of the mobile phase.

$$N = 3482 \ F^{-0.4} \tag{7}$$

Then, a chromatogram, based on the assumption that the peak band is Gaussian, can be drawn according to the following equation-8;

$$y = 1/2 \sqrt{\pi} \sigma \exp(-(x-t_r)^2/ (2 \sigma^2)), \quad \sigma = t_r/\sqrt{N}, \quad t_r-4\sigma \leqq x \leqq t_r+4\sigma \tag{8}$$

where x and y represent time and peak intensity, respectively, and is the standard deviation of the Gaussian curve corresponding to band width, t_w.

The distribution function, y in the equation-8 is normalized in such a way that $\int y dx$ is equal to unity.

An example to demonstrate the performance of this function or the process for the retention prediction is shown in Figure 3 for separation of alkylbenzenes mixture.

Table II.   Capacity factor values at various mobile phase compositions.

(a) acetonitrile/water system

compound	volume fraction of acetonitrile in mobile phase					
	0.70	0.65	0.60	0.50	0.40	0.30
toluene					11.8	
ethylbenzene	1.16		2.06	3.80	22.3	
tert-butylbenzene			6.54			
p-diisopropylbenzene	2.43		5.37	13.1		
indan	1.32		2.35	4.48		
naphthalene					26.0	
acenaphthylene					40.8	
anthracene				9.55		
benzo(j)fluoranthene	5.10		11.4			
aniline	0.56		0.77	1.03	1.53	2.35
nitrobenzene	0.68		1.10	1.75	3.23	6.14
dimethylphthalate	0.59		0.90	1.35	2.49	4.88
o-nitrotoluene	0.80		1.34	2.26	4.69	10.4
p-chloroaniline			1.02	1.52	2.78	5.33
α-bromo-p-nitrotoluene	0.82		1.48	2.74	6.40	16.1
p-nitrophenol		0.49	0.62	0.94	1.65	2.94
p-methoxyphenol		0.42	0.52	0.71	1.07	1.64
p-hydroxyacetanilide		0.27	0.31	0.34	0.47	0.59

(b) methanol/water system

compound	volume fraction of methanol in mobile phase				
	0.70	0.70	0.60	0.50	0.40
benzene					5.21
ethylbenzene			4.35		23.1
m-ethyltoluene	1.39	3.04	7.24	18.9	
1,2,3-trimethylbenzene				17.2	
tert-butylbenzene	1.47	3.47			
indene	0.97	1.82	3.76	8.29	18.6
naphthalene				11.9	30.9
acenaphthene			10.4	29.8	
pyrene			29.2		
benz(a0anthracene	5.04	16.2			
benzo(j)fluoranthene	7.26	25.0			
anisole	0.63	1.05	1.85	3.31	6.25
dimethylbenzene	0.50	0.77	1.24	2.28	4.29
α-bromoacetophenone	0.59	0.93	1.70	3.28	7.05
p-chloroaniline	0.57	0.90	1.46	2.62	4.93
m-aminoacetophenone	0.39	0.52	0.73	1.10	1.87
phenol	0.38	0.54	0.76	1.11	1.79
m-ethylphenol	0.55	0.88	1.55	2.97	6.18
m-aminophenol	0.30	0.39	0.45	0.57	0.77
p-aminophenol	0.79	1.00	1.04	1.32	1.82
o-nitrophenol	0.56	0.90	1.44	2.48	4.39

Table III. Results of multiple regression analysis of the correlation
between log k', P_1 and P_2.

(a) for alkylbenzenes and PAHs, where P_1 is log P and P_2 is F_c, respectively.

| X | mobile phase system | | | | | | | |
| | acetonitrile/water | | | | methanol/water | | | |
	a	b	c	r	a	b	c	r
0.8	–	–	–	–	0.151	0.083	–0.892	0.997
0.7	0.096	0.073	–0.609	1.000	0.228	0.088	–0.873	0.999
0.6	0.180	0.045	–0.487	1.000	0.310	0.091	–0.806	0.998
0.5	0.253	0.038	–0.420	0.999	0.330	0.108	–0.578	0.997
0.4	–	–	–	–	0.369	0.154	–0.538	0.995
0.3	0.362	0.085	–0.208	1.000	–	–	–	–

(b) for polar groups substituted benzenes except phenols, where P_1 is π and P_2 is (HA–HD), respectively.

| X | mobile phase system | | | | | | | |
| | acetonitrile/water | | | | methanol/water | | | |
	a	b	c	r	a	b	c	r
0.8	–	–	–	–	0.108	–0.055	–0.164	0.958
0.7	0.141	–0.074	–0.072	0.986	0.158	–0.073	0.067	0.968
0.6	0.212	–0.082	0.144	0.976	0.212	–0.088	0.325	0.949
0.5	0.303	–0.105	0.378	0.976	0.263	–0.096	0.607	0.955
0.4	0.427	–0.134	0.703	0.986	0.297	–0.105	0.911	0.932
0.3	0.563	–0.164	1.055	0.992	–	–	–	–

(c) for phenols, where P_1 is π and P_2 is $\pi\sigma_h(1-\pi)$, respectively.

| X | mobile phase system | | | | | | | |
| | acetonitrile/water | | | | methanol/water | | | |
	a	b	c	r	a	b	c	r
0.8	–	–	–	–	0.161	0.200	–0.408	0.951
0.7	–	–	–	–	0.204	0.200	–0.238	0.994
0.65	0.156	0.180	–0.401	0.959	–	–	–	–
0.6	0.185	0.180	–0.310	0.962	0.291	0.200	–0.074	0.994
0.5	0.282	0.180	–0.173	0.975	0.361	0.200	0.127	0.992
0.4	0.351	0.180	0.029	0.991	0.430	0.200	0.359	0.991
0.3	0.454	0.180	0.227	0.995	–	–	–	–

Table IV. Retention description for aromatic compounds with the C18 column as a function of organic modifier concentration in mobile phase.

compound group	mobile phase	retention description
alkylbenzenes PAHs	acetonitrile/water	$\log k' = (0.102X^3-0.746X^2+0.427) \log P + (1.022X^2-1.051X +0.308) F_c - (0.980X-0.084)$
alkylbenzenes PAHs	methanol/water	$\log k' = (-0.184X^4-0.327X^3+0.392) \log P + (0.675X^2-0.971X+0.431) F_c + (0.738X^4-1.701X+0.155)$
substituted benzenes	acetonitrile/water	$\log k' = (0.769X^3-1.661X+1.041) \pi - (0.233X^4-0.365X+0.273)(HA-HD) + (2.583X^2-5.405X+2.444)$
substituted benzenes	methanol/water	$\log k' = (0.129X^4-0.505X^2+0.377) \pi + (0.108X^3-0.111) (HA-HD) + (0.678X^3-3.444X+2.245)$
phenols	acetonitrile/water	$\log k' = (0.316X^2-1.150X+0.769) \pi + 0.180 \pi\sigma_h(1-\pi) + (1.117X^2-2.843X+0.981)$
phenols	methanol/water	$\log k' = (0.136X^4-0.824X+0.760) \pi + 0.200 \pi\sigma_h(1-\pi) + (1.145X^2-3.273X+1.482)$

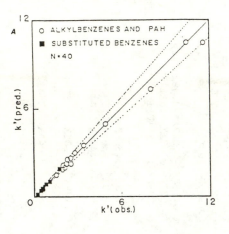

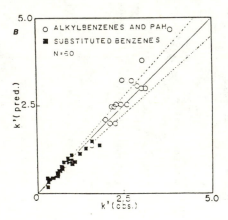

Figure 1.   Relationships between predicted and observed k' values
on the C18 stationary phase.
A : mobile phase; acetonitrile:water = 65 : 35
B : mobile phase; methanol:water = 75 : 25
Reproduced with permission from Ref. 10.   Copyright 1984, Elsevier.

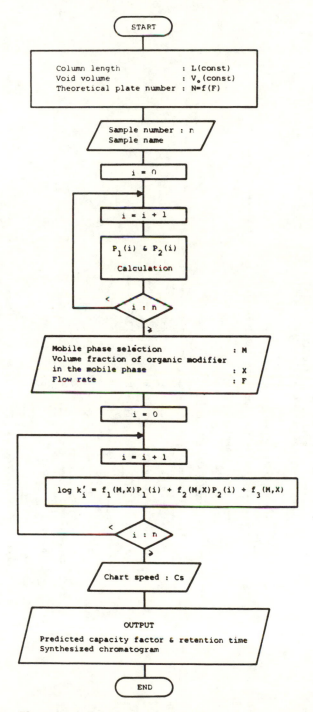

Figure 2.   Flow-chart for retention prediction in reversed-phase LC.   Reproduced with permission from Ref.10.   Copyright 1984, Elsevier.

```
1. synthesizing chromatogram under given conditions
[INPUT]    solutes to be analyzed and separation conditions
[OUTPUT]   chromatogram to be obtained

2. setting up the optimum separation condition
[INPUT]    solutes to be analyzed, analysis time and resol
           tion for two solutes which are of interest
[OUTPUT]   separation conditions optimized in this system
           and the chromatogram under these conditions


please push function number    1

*************************** COMPOUND NAME INPUT ***************************

1. alkylbenzenes (example:C6H5CH3) ? 3
   No. 1   C6H5C2H5
   No. 2   C6H5C(CH3)3
   No. 3   C6H4CH(CH3)2CH(CH3)2

2. polycyclic aromatic hydrocarbons (example:chrysene) ? 0

3. substituted benzenes (example: C6H5NHCH3) ? 0

4. phenols (example:o-methylphenol) ? 0

***** PLEASE SET UP THE SEPARATION CONDITIONS *****

mobile phase (1. CH3CN:H2O 2. CH3OH:H2O)   1
mobile phase composition ( xx : xx )       50  :  50
flow rate (ul/min)                          4

column   : FineSIL C18-5
column dim.: 0.5mm i.d. x 12cm
eluent : CH3CN:H2O = 50 : 50
flow rate: 4 (ul/min)
void time: 5 (min)
theoretical plate number: 2000

*************** RETENTION DATA ***********************
compound                     k'         retention time (min)
C6H5C2H5                    3.77            23.86
C6H5C(CH3)3                 6.84            39.20
C6H4CH(CH3)2CH(CH3)2       13.37            71.87
***********************************************************
please select chart speed
2  (mm/min)
```

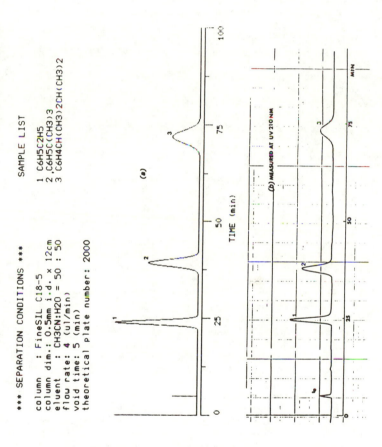

*** SEPARATION CONDITIONS ***   SAMPLE LIST

column   : FineSIL C18-5   1 C6H5C2H5
column dim.: 0.5mm i.d. x 12cm   2.C6H5C(CH3)3
eluent   : CH3CN:H2O = 50 : 50   3.C6H4CH(CH3)2CH(CH3)2
flow rate: 4 (ul/min)
void time: 5 (min)
theoretical plate number: 2000

Figure 3.   Example of the function of RPS in predicting retention
times of three alkylbenzenes.   Input information is underlined.
(a) predicted chromatogram.
(b) measured chromatogram.
Reproduced with permission from Ref.10.   Copyright 1984, Elsevier.

In this example, the chemical formulas of three alkylbenzenes such as ethylbenzene, tert-butylbenzene and p-diisopropylbenzene are input in RPS and then the volume fraction of acetonitrile in the mobile phase is set to 0.50 and the flow rate is 4 μL/min. The chromatogram synthesized by RPS as shown in Figure 3-(a) was compared with that obtained by the actual run as shown in Figure 3-(b). The agreement between both chromatograms is excellent. This result indicates that the system provides useful information for separation conditions.

By the use of the main function of RPS it is possible to predict retention of solutes under any set of conditions. This function will allow determination of optimal separation conditions by evaluation of the separations under various possible combinations of chromatographic parameters.

In order to evaluate the separation, we have to know how resolution (R_S) and analysis time (T_m) vary with experimental parameters such as F, k' and N.

R_S for closely spaced bands can be derived as

$$R_S = 1/4 \; (\; \alpha/\alpha-1)(k_L'/(1+k_L') \sqrt{N}, \quad \alpha=k_L'/k_S', \quad k_L' > k_S' \qquad (9)$$

where k_L' and k_S' are the capacity factors for bands L and S, respectively, and α is the separation factor and the theoretical plate number, N is determined by the equation-7 as the function of flow rate F.

Time required to complete the separation is calculated by the equation-10 and 11;

$$T_m = t_{rmax} + 4 \; \sigma_{max}, \quad \sigma_{max} = t_{rmax} / \sqrt{N} \qquad (10),(11)$$

where t_{rmax} is the retention time of the last eluted solute.

The flow-chart of this function in RPS is shown in Figure 4. When RPS is used as the system to obtain an optimized separation condition, the following data required:
(1) the compound name or the chemical formula of interesting solutes,
(2) the analysis time requested (T_m^*),
(3) the resolution for the two solutes of interest (R_S^*).

By input compounds' names or chemical formulas to RPS, suitable descriptors for the compounds' group desired are calculated with the same procedures as in the main function of RPS by the computer and then capacity factors for the solutes at various mobile phase compositions are predicted by step-by-step with the interval of X=0.01 for both aqueous acetonitrile and methanol mobile phases. The range available in this procedure is from 0.3 to 0.7 of X-values for acetonitrile system and from 0.4 to 0.8 for methanol system, respectively. After calculations of capacity factors for the desired solutes, R_S and T_m for each step are estimated according to the equation-9 and 10, at five different flow rates of the mobile phase such as 1, 2, 4, 8 and 16 μL/min (because we use microcolumns) and then quality of the separation is judged using a simple numerical chromatographic response function (CRF) defined as follows:

$$CRF = | \; T_m - T_m^* \; | / \; T_m^* + | \; R_S - R_S^* \; | / \; R_S^* \qquad (12)$$

The most optimized conditions are obtained when the CRF value is the minimum. In this definition of CRF, if R_S is too large compared

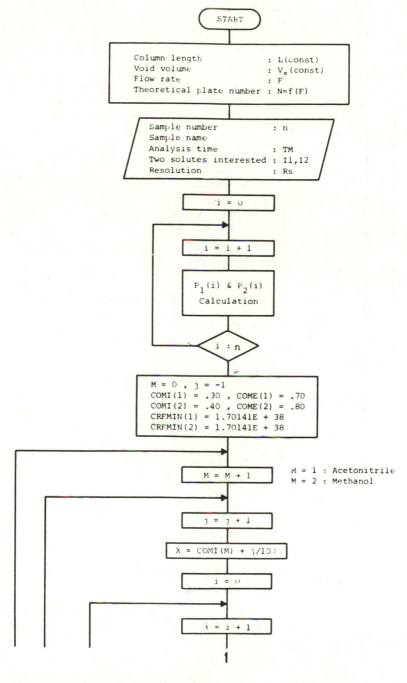

Figure 4. Flow-chart for the optimization of separation conditions in reversed-phase LC. Reproduced with permission from Ref.
9. Copyright 1984, Elsevier.

Part 1 of 3. Continued on page 182.

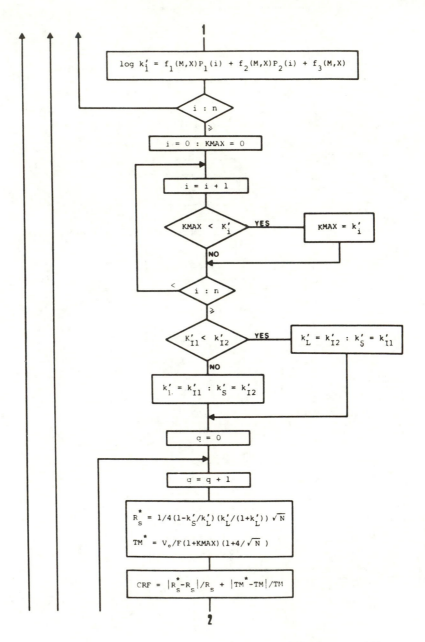

Figure 4.   Part 2 of 3.   Continued on page 183.

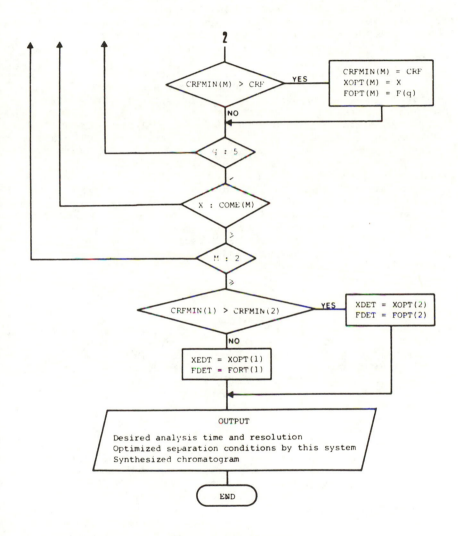

Figure 4.   Part 3 of 3.

with the requested value, it is considered as the excessive condition
and the system searches the condition closer to the requested one.
At this stage, the user can prioritize resolution or analysis time.
In case the priority is resolution, the first term of the equation-12
should be weighted to zero, while the second term should be weighted
to zero in the opposite case.

    With over 800 times calculations, RPS can find the optimized se-
paration condition for the desired solutes and the system provides
numerical values such as mobile phase composition, flow rate, analysis
time and resolution attained at that condition, and then RPS draws
the idealized chromatogram at the optimized separation condition on
the CRT.

    To demonstrate the potential of this function of RPS, an experi-
ment was performed. The demonstration as shown in Figure 5 is the
separation of a test-mixture of aniline derivatives. The chemical
formulas of three aniline derivatives such as aniline, N-methylaniline
and N-ethylaniline were input in RPS and the analysis time of 10 min
and a resolution value of 1.2 for N-methyl- and N-ethylanilines were
requested. RPS provided the predicted retention data of above three
compounds; both the conditions desired by the user and conditions
attained in this system. The system showed that analysis time of 11
min and the resolution of 1.1 would be obtained at the 65 % acetonitrile in the mobile phase and the flow rate of 4 µL/min. The synthesized chromatogram appeared on the CRT and the printer connected
with the computer. The actual run was carried out to evaluate the
performance of RPS at that optimized separation condition of 65 %
acetonitrile in the mobile phase and the flow rate of 4 µL/min. The
system's analysis time of 11 min compares favorably with measured
analysis time of 11.5 min.

Conclusion

The procedure to construct RPS for reversed-phase LC was established,
based on the QSRR studies. In this procedure, if the information
about retention data of several standard materials at some mobile
phase compositions (four or five) for a given column is input into the
computer, it is possible to predict the retention of the solutes,
which belong to the same class of the compounds, with fairly good
accuracy in a few second, a relative error of $\pm$ 10 % for the measured
retention data.

    The accuracy of this system is dependent on the correlation coe-
fficient of a retention description obtained from studies of QSRR,
therefore, the selection of descriptors is the most basic and important task to construct RPS. This selection could be done with statistical framework, even if such description is not clearly derived from
theories. The retention description obtained from QSRR studies is
more effective for a rapid and accurate prediction of retention than
that derived from theoretical models, because the former is simple
and does not require introduction of a number of physicochemical parameters (they are often not clearly known and are very difficult and
time-consuming to determine) for the latter case. By contrast, the
consideration of physical meanings of descriptors derived from QSRR
studies gave the overview of retention mechanisms in reversed-phase
LC (7-10). That is to say, hydrophobicity, size and shape of alkylbenzenes and PAHs are dominate factors controlling their retention,

```
1. analysis time (min) ?  10           SAMPLE LIST
2. two solutes' number of inter-
   est ?  2,3                          No. 1   C6H5NH2
   resolution for these compounds      No. 2   C6H5NHCH3
   to be desired ?  1.2                No. 3   C6H5NHC2H5


column     : FineSIL C18-5
column dim.: 0.5mm i.d. x 12 cm
eluent     : CH3CN:H2O = 65 : 35
flow rate: 4 (ul/min)
void time: 5 (min)
theoretical plate number: 2000

result output (Y/N) ? Y


*********************** RETENTION DATA ***********************
compound                    k'              retention time (min)
C6H5NH2                     0.65                8.26
C6H5NHCH3                   0.88                9.41
C6H5NHC2H5                  1.10               10.48
************************************************************

***************** CONDITIONS DESIRED BY USER *****************
analysis time : 10 (min)
solutes of interest : C6H5NHCH3  &  C6H5NHC2H5
resolution of these solutes to be desired : 1.2
************************************************************

************* CONDITIONS ATTAINED IN THIS SYSTEM ************
analysis time : 11 (min)
resolution to be attained : 1.1
************************************************************

*** SEPARATION CONDITIONS ***                SAMPLE LIST

column     : FineSIL C18-5                   1 C6H5NH2
column dim.: 0.5mm i.d. x 12 cm              2 C6H5NHCH3
eluent     : CH3CN:H2O = 65 : 35             3 C6H5NHC2H5
flow rate: 4 (ul/min)
void time: 5 (min)
theoretical plate number: 2000
```

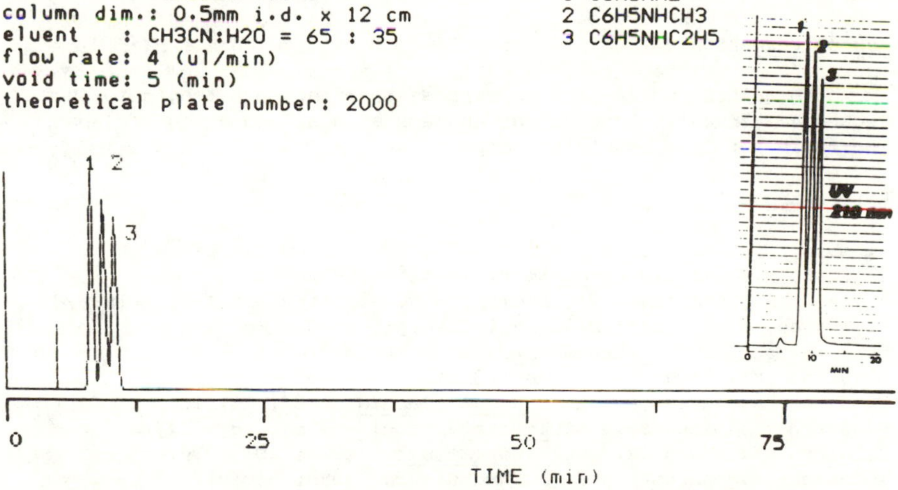

Figure 5. Example of the optimization of the separation of aniline
derivatives. Input information is underlined. Reproduced with
permission from Ref. 9. Copyright 1984, Elsevier.

while for the polar groups substituted benzenes, hydrophobicity and electronic properties of substituents are factors to govern their retention in reversed-phase LC.

The function of RPS can aid setting up the chromatographic separation conditions. The optimization of a separation can be done experimentally based a chromatographer's experience and intuition, or with computer control based on the "Chemometrics", i.e., "Simplex" procedure (4,5). Although such approaches are reliable and promising enough to use for practical purposes, this is a time-consuming task because a number of experiments required with the increase factors for optimization. On the other hand, it is possible to optimize a separation in RPS with almost the same accuracy as obtained in the conventional Chemometrics approaches within a few minutes.

RPS constructed here can be used for the positive identification and gives the information for setting up the chromatographic conditions. Further accomplishments on the following items will make the system more general and increase the applicability of RPS:
(1) the introduction of more universal expressions for retention description of any compounds types and columns.
(2) to apply this system for gradient elution mode and/or for ternary mobile phase systems.
(3) on-line prediction of solute retention and optimization of LC-separations.

In addition, it is apparent from the results described in this communication that the rapid identification of the solute is possible without any standard materials. Focusing on some selected compounds identified by the method here, more precise information could be obtained by the more advanced and complicated analytical techniques such as GC-MS with spectra data-base system. Naturally, the other spectroscopic techniques such as FTIR or UV (with photodiode array for three dimensional measurements) (43,44) combined with the system might have an even higher power in the identification processes of practical samples.

Acknowledgments

The authors would like to express their sincere thanks to The Mechanical Industry Development and Assistance Foundation of Japan for their partly financially support.

Literature Cited

1. Lamb, R.J.   Amer.Lab. 1981, 13(3), 47.
2. Lamb, R.J.   J.Liq.Chromatogr. 1984, 7, 647.
3. Sachok, B.; Korg, R.C.; Deming, S.N. J.Chromatogr. 1980, 199, 317.
4. Watson, M.W.; Carr, P.W. Anal.Chem. 1979, 51, 1835.
5. Berridge, J.C. J.Chromatogr. 1982, 244, 1.
6. Jinno, K.; Kawasaki, K. Anal.Chim.Acta 1983, 152, 25.
7. Jinno, K.; Kawasaki, K. Chromatographia 1984, 18, 103.
8. Jinno, K.; Kawasaki, K. Chromatographia 1984, 18, 211.
9. Jinno, K.; Kawasaki, K. J.Chromatogr. 1984, 298, 326.
10.Jinno, K.; Kawasaki, K. J.Chromatogr. 1984, 316, 1.
11.Yonker, C.R.; Zwier, T.A.; Burke, M.F. J.Chromatogr. 1982, 241,257.
12. Yonker, C.R.;Zwier, T.A.; Burke, M.F. J.Chromatogr. 1982, 241,269.
13. Karger, B.L.; Gant, J.R.; Hartkoph, A.; Weiner, P.H. J.Chromatogr.
    1976, 128, 65.

14. Hemetzberger, H.; Behrensmeyer, P.; Hemming, J.; Ricken, H. Chromatographia 1979, 12, 71.
15. Kikta, E.J. Jr.; Grushka, E. Anal.Chem. 1976, 48, 1098.
16. Sander, L.C.; Callis, J.B.; Field, L.R. Anal.Chem. 1983, 55,1068.
17. Gilpin, R.H.; Gangoda, M.E. J.Chromatogr.Sci. 1983, 21, 352.
18. Miller, M.L.; Linton, R.W.; Bush, S.G.; Jorgenson, J.W. Anal.Chem. 1984, 56, 2204.
19. Horvath, C.; Melander, W.; Molnar, I. J.Chromatogr. 1976, 125, 129.
20. Melander, W.; Stoveken, J.; Horvath, C. J.Chromatogr. 1980, 199, 35.
21. Horvath, C.; Melander, W. Amer.Lab. 1978, 10(10), 17.
22. Mirrless, M.S.; Moulton, S.J.; Marphy, C.T.; Taylor, P.J. J.Med. Chem. 1976, 19, 615.
23. Konemann, H.; Zelle, R.; Busser, F.; Hammers, W.E. J.Chromatogr. 1979, 178, 559.
24. Miyake, K.; Terada, T. J.Chromatogr. 1982, 240, 9.
25. Lins, C.L.K.; Block, J.H.; Doerge, R.F.; Barnes, G.L. J.Pharm.Sci. 1982, 71, 614.
26. Jinno, K. Chromatographia 1982, 15, 667.
27. Biagi, G.L.; Barbaro, A.M.; Gueira, M.C.; Babbini, M.; Gaiardi, M. ; Bartoletti, M.; Borea, P.A. J.Med.Chem. 1980, 23, 193.
28. Jinno, K. Anal.Lett. 1982, 15(A19), 1533.
29. Melander, W.R.; Horvath, Cs. Chromatographia 1982, 15, 86.
30. Knox, J.H., Kaliszan, R.; Kennedy, G.J. In"Chromatography, Equi- and Kinetics", Faraday Symposia of the Chemical Society, 1980, 113.
31. McCormick, R.M.; Karger, B.L. Anal.Chem. 1980, 52, 2249.
32. Wells, M.J.M.; Clark, C.R. Anal.Chem. 1981, 53, 1341.
33. Berendson, G.E.; Schoemakers, P.J.; de Galan, L.; Vigh, G.; Varga Puchong, Z.; Inczedy, J. J.Liq.Chromatogr. 1980, 3, 1669.
34. Grushka, E.; Colin, H.; Guiochon, G.G. J.Liq.Chromatogr. 1982, 5, 1391.
35. Grushka, E.; Colin, H.; Guiochon, G.G. J.Chromatogr. 1982, 248, 325.
36. Jinno, K.; Ozaki, N.; Sato, T. Chromatographia 1983, 17, 341.
37. Jinno, K. Chromatographia 1983, 17, 367.
38. Rekker, R.F. "The Hydrophobic Fragmental Constant"; Elsevier: Amsterdam, 1977.
39. Hansch, C.; Leo, A. "Substituent Constants for Correlation Analysis in Chemistry and Biology"; Wiley Intersci.: New York, 1979.
40. Bondi, A.   J.Phys.Chem. 1964, 68, 441.
41. Schabron, J.F.; Hurtubise, R.J.; Silver, H.F. Anal.Chem. 1977, 49, 2253.
42. Hurtubise, R.J.; Allen T.W.; Silver, H.F. J.Chromatogr. 1982, 235, 517.
43. Jinno, K.; Fujimoto, C.; Uematsu, G. Amer.Lab. 1984, 16(2), 39.
44. Fell, A.F.; Clark, B.J.; Scott, H.P. J.Pharm.Biomed.Anal. 1983, 1, 557.

RECEIVED May 2, 1985

11

Calculation of Retention for Complex Gradient Elution High-Performance Liquid Chromatographic Experiments: A Universal Approach

Sterling A. Tomellini[1,3], Shih-Hsien Hsu[1], R. A. Hartwick[1], and Hugh B. Woodruff[2]

[1] Department of Chemistry, Rutgers, The State University of New Jersey, New Brunswick, NJ 08903
[2] Merck Sharp & Dohme Research Laboratories, Rahway, NJ 07065

A numerical method for the calculation of retention under complex gradient HPLC elution conditions is presented. The approach is applicable to virtually any gradient or solvent-solute relationship. Examples include binary, ternary, quaternary, linear and multiple-linear solvent gradients as well as for cases involving stationary phase programming (coupled column gradient elution experiments).

The chromatographic experiment is somewhat unique in that more time is normally spent designing the experiment than interpreting the resulting data. It is for this reason that much of the current work in liquid chromatography is directed at helping the researcher determine which experimental conditions will produce an adequate separation. A number of papers have been published which demonstrate various approaches for the "optimization" of liquid chromatographic parameters. The researcher should, however, be conscious of two questions before choosing to use any "optimization" strategy. First, one should ask when an optimization strategy should be used. The second question the researcher should address is what is meant by an "optimum" in regard to any LC separation and in particular to the separation which is presently under consideration.

It is obvious that the only reason to use an optimization strategy is that the response surfaces are unknown and would require too much effort to evaluate directly. Liquid chromatography has an added complication in that the chromatographic response may be time dependent (due to changes in the column, etc.). The problem with all such strategies is, however, defining what is meant by the "optimum" separation. The problems associated with such indicators are well known (1). DeGalan (2) has demonstrated that the chromatographic response surface actually consists of two independent surfaces (time and resolution). The "optimum" separation will, therefore, most

[3] Current address: Department of Chemistry, University of New Hampshire, Durham, NH 03824

probably be a compromise between time and resolution. Furthermore, it should be emphasized that the information desired from a chromatographic experiment must also be considered when determining the "optimum" separation. Thus, the "optimum" separation can vary considerably among various researchers even for a single sample.

An alternative approach for determining which conditions produce adequate separations is to take a limited amount of chromatographic data and use available knowledge to predict expected responses. Such an approach can be termed a "calculational" approach and can generally be broken into two parts. The available data must first be fitted to an expected retention vs. solvent concentration curve. Once the coefficients for this curve are known, calculations can be made for both isocratic and gradient conditions. The gradient equations can be solved using either numerical integration techniques or exact solutions. Normally exact calculations can be used to predict elution times and widths for isocratic conditions. Exact solutions are also available for some gradient conditions (3-21); the solutions, however, are dependent on gradient shape, number of solvent components and the relationship between solvent components and retention data. Often exact solutions are not available for complex gradient conditions which leads to the use of more flexible techniques employing numerical integration (22). Furthermore, use of numerical methods to treat gradient conditions allows for independence between the fitting of retention to solvent concentration data and the gradient calculations. Also, numerical integration is generally more computer amenable as programming complexity is significantly reduced if only one calculational algorithm is required.

The development of computer programs capable of using the calculation approach to assist the researcher in determining which chromatographic conditions are useful will be presented. The theory, assumptions, advantages and disadvantages of the calculations on which the programs are based will also be presented. Examples will be given for separations of test compounds using binary, ternary and quanternary, linear and multiple-linear gradients. Calculation of retention times for combined stationary and mobile phase programming will also be presented.

Theory

Single Column. It is well known that the capacity factor, k', for a compound often varies in a predictable manner with respect to the concentration of organic modifier for aqueous solutions under reverse phase conditions. It has been shown for a single organic modifier the relationship is:

$$Ln(k') = AC_1 + DC_1{}^2 + B \qquad (1)$$

where A, B, and D are constants and C_1 is the concentration of organic modifier which is generally given in volume percent. Additional modifiers are expected to affect the capacity factor of a compound in a similar manner. It is expected, therefore, that

for an aqueous solution containing two organic modifiers the
relationship will be:

$$Ln(k') = AC_1 + BC_1{}^2 + DC_2 + EC_2{}^2 + FC_1C_2 + G \qquad (2)$$

where A, B, D, E, F and G are constants. Notice that a term has
been included which varies as a function of both C_1 and C_2. Using
similar mathematical fitting any number of organic modifiers can be
used.

Once the expected relationship between k' and organic modifer
is known, the values of the constant coefficients must be
determined. Using a limited amount of experimentally determined
isocratic data the coefficients can be calculated by using a linear
least squares fit algorithm. The minimum number of data points
required varies with the number of terms in the equation being fit.
The data for mobile phases containing one organic modifier are fit
to Equation 1 and require at least 3 experimentally determined data
points. Likewise, 6 data points are required for aqueous mobile
phases having two organic modifiers (e.g. aqueous solution
containing methanol and acetonitrile as modifiers). Using similar
fitting equations any number of organic modifiers may be employed.
Liquid chromatography is often thought of as a volume problem but
is in fact fundamentally a length problem since the column length
is a limiting parameter. All separations must be completed by the
time the compounds have traveled the length of the column. Thus,
one of the fundamental equations of LC is:

$$Length = Velocity \times Time \qquad (3)$$

In the isocratic case the velocity is constant and Equation 3
becomes:

$$L_{col} = t_R \times U_{band} \qquad (4)$$

Where L_{col} is the column length, t_R is the retention time for a
given solute and U_{band} is the linear velocity of that solute band.
If the velocity is not constant, as is generally the case in
gradient elution, the integral solution for Equation 4 must be used
which is given as:

$$L_{col} = \int_0^{t_R} U_{band.inst} \cdot dt \qquad (5)$$

where $U_{band.inst}$ is the instantaneous velocity of the solute band
and the integration limits are from time = 0 to time = t_R.

The velocity of a band can be found for the isocratic case by
first rearranging the often used expression for k':

$$k' = (t_R - t_M)/t_M \qquad (6)$$

to get

$$t_R = t_M(1 + k') \qquad (7)$$

where t_M is the elution time of an unretained solute. U_{band} for elution under isocratic conditions can be calculated by dividing the column length by t_R (as given in Equation 7) resulting in:

$$U_{band} = L_{col}/(t_M(1 + k')) \tag{8}$$

For the gradient case where k' is generally changing throughout, the instantaneous band velocity is of interest. It follows from Equation 8 that $U_{band.inst}$ can be found in terms of the instantaneous k', k'_{inst}, such that:

$$U_{band.inst} = L_{col}/(t_M(1 + k'_{inst})) \tag{9}$$

Thus the fundamental equation for gradient elution becomes:

$$L_{col} = \int_0^{t_R} L_{col}/(t_M(1 + k'_{inst})) \cdot dt \tag{10}$$

Obviously, Equation 10 reduces to Equation 4 for isocratic cases where k'_{inst} is constant over time.

Using Equation 10 it is, in theory, possible to solve exactly for the retention time of any band knowing the relationship between the solute's instantaneous k' and time. If, however, the instantaneous k' vs time relationship is mathematically complex, as can be the case in gradient elution, then an exact solution for t_R will be difficult if not impossible to determine. There are a number of factors affecting the complexity of the k'_{inst} vs. time relationship, among these are:

1. The number of modifiers in the mobile phase and the relationship between k' and concentration for these modifiers.
2. The gradient shape (i.e., linear, multiple linear, complex).
3. The instrumental delay time which causes most solutes to travel isocratically before being overtaken by the gradient.
4. A time correction which is necessary since the gradient front after overtaking the solute band generally moves at a faster velocity than the solute band.

While exact mathematical solutions for retention times are preferable (when available) for manual calculations, numerical integration has two major advantages when the computing power of a large computer is available. First, solutions are not always available. Second, the versatility of numerical integration reduces the amount of programming effort necessary and also the size of the required programs if multiple experimental conditions are to be allowed.

As when using exact solutions, the first step when using solutions employing numerical integration is to determine the relationship between k' and solvent composition for each solute using limited isocratic data. Next the retention times of each solute must be calculated in turn. First, the time spent by the

solute traveling under the initial isocratic conditions caused by
the instrumental delay time, t_D, must be calculated. Knowing the
actual time delay before the gradient front reaches the solute band
and the k' of the solute band during this time, the distance
traveled by the solute under isocratic conditions can be
determined.   Notice that this problem is essentially a related
rate problem of the type normally presented in beginning calculus
courses.   It is easiest to visualize the problem by thinking of the
solute band as having an effective head start in distance equal to
the measured gradient time delay multiplied by the linear velocity
of the gradient front down the column which is equal to L_{col}/t_M.
If the gradient front travels for a time, t_{corr} at a velocity equal
to L_{col}/t_M and the solute band travels for the same time at a
velocity given by Equation 8, $L_{col}/(t_M(1 + k'))$, then knowing the
effective head start of the solute band allows for the calculation
of t_{corr}. Simply stated the distance traveled by the solute band
plus the head start must be equal to the distance traveled by the
gradient front at the time the two are coincidental, or:

$$(L_{col}/t_M(1 + k')) \times t_{corr}) + ((L_{col}/t_M) \times t_D = L_{col}/t_M \times t_{corr} \quad (11)$$

Solving Equation (11) for t_{corr} gives:

$$t_{corr} = ((1 + k')/k') \times t_D \quad (12)$$

The distance traveled by the solute band isocratically, L_{iso} can be
found by substituting Equation 12 into Equation 8 and rearranging
to get:

$$L_{iso} = ((L_{col} \times t_D)/(t_M \times k')) \quad (13)$$

Knowing the time spent by the solute band traveling isocratically
and its position in the column when the gradient takes effect
allows the calculation of the distance over which the gradient will
affect the solute since:

$$L_{iso} + L_{grad} = L_{col} \quad (14)$$

This, L_{grad} can be substituted into Equation 10 for L_{col} to become:

$$L_{grad} = \int_{0}^{t_R'} u/(k'_{inst} + 1) \cdot dt \quad (15)$$

where

$$t_R = t_{corr} + t_R' \quad (16)$$

and u is the linear velocity which is equal to L_{col}/t_M. Notice
that while the problems previously noted as being associated with
the instrumental delay have been overcome, it is still often not
possible to solve Equation 15 exactly. It is easiest, therefore,
to evaluate the integral in a stepwise manner by simply
incrementing the time by some small step, calculating the
corresponding k' for the resulting time and then calculating the

length traveled during the time interval. The sum of the time steps will equal t_R, when the sum of the lengths traveled is equal to L_{grad}. A continuous correction must be made, however, since the actual time spent traveling and the time corresponding to the gradient concentration which the solute band encounters are generally not the same. The reason for this is that the gradient concentration seen by the band is not only dependent on the time after the gradient has started but also on the position of the band in the column. The simplest way to eliminate length from the problem is to calculate a "gradient" time as well as an actual time. While the actual time interval is specified, the "gradient" time interval, $t_{grad.int}$, can be calculated using the relationship:

$$t_{grad.int} = t_{actual.int} - L_{trav.act.int}/u \qquad (17)$$

where $L_{trav.act.int}$ is the length traveled by the solute band in the actual time interval. Thus, the sum of t_{grad} intervals is equal to the "gradient" time seen by the solute band and it is this time that must be used to calculate the instantaneous capacity factors for the stepwise integration.

Multiple Columns. The basic calculational steps are the same for the coupled column experiment as for the single column experiment with the exception of the determination of solute band migration caused by the instrumental gradient delay. The distance a solute band migrates isocratically will depend not only on the mobile phase concentration during the gradient delay but also on the capacity factor vs mobile phase relationship for the solute on each column encountered. The distance migrated isocratically will also depend on the length of each column traversed. Solving the coupled-column gradient experiment thus becomes a problem of tracking throughout the elution the position and speed of all solute bands and the gradient front.

The key to understanding the effect of multiple columns on the time that the gradient overtakes the solute band is to treat each column as a separate delay volume. If the gradient overtakes the band in the first column, then the previously derived relationship between instrumental delay time and the length traveled by the band in the column isocratically, L_{iso}, can be used:

$$L_{iso} = (L_{col} \times t_D)/(t_M \times k') \qquad (18)$$

where L_{col} is the column length for column 1, t_D is the instrumental delay time, t_M is the retention time of an unretained solute for column 1 and k' is the capacity factor at the mobile phase conditions of the instrumental delay for the solute of interest. If, however, the band travels through the first column isocratically then the distance the solute traveled in the second column before being overtaken by the gradient must be calculated. If one considers the situation just as the solute band reaches the head of the second column, then it can be seen that the first column has simply acted as an additional delay volume and an equation similar to Equation 1 can be derived. The correction that must be made is simply that the delay time used for column 2, t_{D2},

is equal to the instrumental delay time plus the retention time of
an unretained solute for column 1, t_{M1} minus the elution time of
the solute band from column 1, t_{R1}. If the elution time of the
band in column 1 is derived in terms of the solute's capacity
factor in column 1, then the delay time for column 2 becomes:

$$t_{D2} = (t_D + (1 + k'_1)(L_{col(1)}/u_1) + t_{M1}) \qquad (19)$$

The same logic can be applied to any number of columns coupled
together and this forms the basis of the developed programs.

Once the position at which the gradient overtakes the solute
band is known, numerical integration allows the calculation of the
solute's retention under gradient conditions. The calculations
involved for the gradient coupled column experiment are essentially
identical to that used for single column gradient elution once the
distance traveled isocratically is known. The integral for the
gradient is evaluated in a stepwise manner by incrementing the time
by some small step (0.01 minute was used for this study),
calculating the corresponding k' for the resulting time and then
calculating the length traveled during the time interval. The sum
of the time steps will equal t_R, when the sum of the lengths
traveled is equal to L_{grad}. A continuous correction must be made,
however, since the actual time spent traveling and the time
corresponding to the gradient concentration which the solute band
encounters are generally not the same. The reason for this is that
the gradient concentration seen by the band is not only dependent
on the time after the gradient has started but also on the position
of the band in the column. As previously stated the easiest way to
eliminate length from the problem is to calculate a "gradient"
time as well as an actual time. While the actual time interval is
specified, the "gradient" time interval, $t_{grad.int}$, can be
calculated using the relationship,

$$t_{grad.int} = t_{actual.int} - L_{trav.act.int}/u_{col} \qquad (20)$$

Where $L_{trav.act.int}$ is the length traveled by the solute band
in the actual time interval, and u_{col} is the linear velocity of the
gradient front in the column presently containing the solute band.
Thus, the sum of $t_{grad.int}$ intervals is equal to the "gradient"
time seen by the solute band and it is this time that must be used
to calculate the instantaneous capacity factors for the stepwise
integration. As in the case of the single column experiment, the
actual retention time for the solute will equal the sum of the time
spent traveling isocratically and the time spent traveling under
gradient conditions.

EXPERIMENTAL SECTION

Single Column Experiments

Apparatus and Reagents. Five test solutes were used to determine
the accuracy of the predictive capabilities of the developed
program. They were: uracil, phenol, acetophenone, nitrobenzene and
methylbenzoate. All compounds used were reagent grade or better.

HPLC grade (J.T. Baker, Inc., Phillipsburg, N.J.) methanol,
acetonitrile and tetrahydrofuran were used as organic modifiers.
Mobile phases were prepared volumetrically using double-distilled,
deionized water and were sparged with helium before use.
     The chromatographic system consisted of a Brownlee MPLC
MicroPump (Brownlee Labs., Santa Clara, Calif.) using a nominal 150
ul packed-bed mixing chamber to ensure the reproducibility of the
gradient concentration, a Rheodyne variable sample loop injection
valve (Rheodyne, Cotati, Calif.) model 7413, set at 0.5 ul and a
Kratos (Kratos Analytical Instruments, Ramsey, N.J.) model SF 769
U.V. detector operating at 254 nm. with a 0.5 ul flow cell. The
analog output of the detector was recorded using a Kipp & Zenon,
BD-40 series, strip chart recorder.
     Laboratory-bonded octyl stationary phases were prepared using
either Whatman Partisil-10 10 um silica gel (batch no. 100591)
(Whatman, Inc., Clifton, N.J.) or Shandon MOS-Hypersil 5 um
spherical support (batch no. 10/899) (Shandon Products, Ltd.,
Cheshire, England). Both phases were bonded by refluxing 3 g of
silica gel with a 5-fold excess of chlorodimethyloctylsilane in
toluene, using 5 ml of pyridine as an acid scavenger. Reaction was
allowed to proceed at 65-70°C for 24 hours. Phases were then
washed in methanol, dried and exhaustively end capped with
trimethylchlorosilane. Stationary phases were packed into 250 x 1
mm id glass lined columns by slurrying 300 mg of packing into 3 ml
of isopropyl alcohol, and packing at a pressure of 10,000 psi for
approximately 15 minutes, using acetone as the packing solvent
(23). The MOS-Hypersil bonded phase was used for the binary
gradient studies while the Partisil-10 bonded phase was used for
ternary and quaternary gradient experiments.

Multiple Column Experiments. Six test solutes were used for the
multiple column experiments. They were: uracil, guaifenesin,
p-cresol, p-ethylphenol, nitrobenzene and methylbenzonate. All
solutes were reagent grade or better. Mobile phases were prepared
as previously stated for single column experiments. The stationary
phases used were produced by reacting the appropriate chlorosilyl
derivative with Whatman Partisil-10. The phenyl phase and C_{18}
stationary phases used were slurry packed into 150 x 1 mm and 100 x
1 mm id glass lined columns, respectively. The chromatographic
system was the same as described previously except a 220 ul
instrumental delay was used for the multiple column experiments.
     All computer programs were written in FORTRAN using a Digital
Equipment Corporation (Maynard, Mass.) VAX 11/780. All
calculations were made using the revised version of a previously
described program (22).

Procedure. Since the MPLC MicroPump is a dual-syringe pump only
two solvent concentrations were used for ternary and quanternary
gradients. Solvent proportioning was made by the pump in all cases
to obtain the required isocratic data. The flow rate throughout
the entire study was kept constant at 50 ul/min. The instrumental
delay volume was varied by injecting the solutes at various times
before and after the gradient was started.

<u>RESULTS AND DISCUSSION</u>

<u>Single Column Experiments</u>

<u>Isocratic Data</u>. As previously described, mobile phases containing
one modifier require at lease three isocratic data points to allow
fitting the expected k' vs modifier concentration curve. Methanol
was used as an organic modifier to test the capabilities of the
program on binary gradients. The retention times of the five test
solutes were determined under four isocratic conditions (35%, 45%,
55% and 65% methanol). The retention data are given in Table I.
Retention times were generally found to vary by less than 0.2
minutes between runs for a given isocratic solvent concentration.
Given the excellent reproducibility of retention times observed,

Table I. Binary Isocratic Data

Point #	1	2	3	4
Solute	Retention Time (Minutes)			
Uracil	3.1	3.2	3.2	3.2
Phenol	4.1	5.2	6.9	10.4
Acetophenone	4.8	6.4	10.2	18.6
Nitrobenzene	5.4	7.4	12.0	20.6
Methylbenzoate	6.2	9.7	18.9	41.4

Mobile Phase Concentration

Point #	% Methanol
1	65.0
2	55.0
3	45.0
4	35.0

a. All times are in minutes.
b. All solvent compositions are in volume percent.

single data points at any given solvent composition were
subsequently used for the fitting of k' vs solvent composition
data. Multiple experiments could be performed for each composition
if desired, however, in practice this was not found to be
necessary.

<u>Binary Gradients</u>. Retention data for a number of binary gradients
were acquired and compared to predicted values. The instrumental
delay volume was determined to be 155 ul. The actual and
calculated retention times (along with the percent differences) for
a 10 minute gradient from 35% to 65% methanol are given in Table
IIa. The same gradient was run with a 255 ul instrumental delay
simulated by injecting the solutes 100 ul (2.0 minutes) before
starting the gradient. The retention data for this gradient are
also given in Table IIa. The program easily accommodated the

varying instrumental delays, allowing the scientist to evaluate quickly the effects of such a delay.

Data were also acquired for binary multiple linear gradients. The actual and calculated retention data for one of these gradients are presented in Table IIb. In this instance, only the last two compounds, nitrobenzene and methylbenzoate, ever experienced the second segment of the gradients. The actual and calculated retention values are again in excellent agreement, demonstrating the accuracy of both the mathematical approach and program. Total error in both the generation of the gradient and computer calculations averaged better than 1.5% for the demonstrated cases.

Ternary Gradient Systems. While the fitting of the isocratic concentration vs retention relationship is fairly straightforward for mobile phases containing one modifier, it becomes increasingly difficult as the number of modifiers increases. Ternary solvent systems contain two independent and one dependent variables and require at least six isocratic data points to fit to Equation 2.

Theoretically, a well designed multivariate experiment should be used for determining the capacity factor response surface for ternary and higher order systems. Experimentally, however, such multivariate experiments can be extremely difficult to perform due to instrumental and time limitations. To demonstrate the approach of numerical integration for complicated relationships between capacity factor and solvent composition, as are typically encountered in ternary and quaternary chromatographic systems, it was decided to acquire data along a solvent "line" generated by allowing the binary HPLC pump to proportion two ternary (or higher) reservoir solvents to obtain the required number of isocratic data points. These isocratic data were then fit to expected k' vs solvent composition relationships. Such "line" experiments have inherent mathematical limitations and instabilities of which the user must be aware. A fit is being forced to a given relationship and this relationship may have no physical meaning, therefore, useful results ,if any, can only be expected within the chosen boundary conditions. Furthermore, good fits may exist between other, possibly less complicated relationships and the experimental data. However, the use of a solvent "line" represented a practical means to simulate the complexity of a full surface map utilizing the available instrumentation.

Ternary gradients were run using aqueous solutions containing methanol and acetonitrile as organic modifiers. A solution of 50:10:40 acetonitrile:methanol:water was used as the high strength solvent while the low strength solvent was a mixture of 5:30:65 acetonitrile:methanol:water. The pump itself was used to proportion these two solvents for the isocratic experiments. The ternary isocratic retention data are given in Table III. Figure 1 illustrates by use of a ternary diagram, both the solvent compositions which are possible using the solvents chosen for the reservoirs and also the points at which isocratic data were acquired.

Using the fitted ternary data, a ten minute gradient was run from 14.0% acetonitrile, 26.0% methanol to 41.0% acetonitrile, 14.0% methanol. The gradient was run with a 155 ul instrumental delay, producing the chromatogram shown in Figure 2. The

Table II. Binary Gradient Data

a. Single Segment Linear Binary Gradient Profile

Conditions:
    Solvents: Methanol/Water
    Gradient: 35.0:65.0 $CH_3OH:H_2O$
       to   65.0:35.0 $CH_3OH:H_2O$
              over 10 min.
----

1. 155 ul Instrumental delay

Solute	Obs. t_R	Calc. t_R	% diff.
Uracil	3.2	---	---
Phenol	9.2	9.3	1.1
Acetophenone	12.2	12.1	0.8
Nitrobenzene	13.0	12.9	0.8
Methylbenzonate	15.4	15.2	1.2

2. 255 ul Instrumental delay

Solute	Obs. t_R	Calc. t_R	% diff.
Uracil	3.2	---	---
Phenol	9.6	10.0	4.2
Acetophenone	13.4	13.5	0.7
Nitrobenzene	14.4	14.4	0
Methylbenzoate	17.2	16.9	1.7

b. Multiple Linear Binary Gradient Profile

Conditions:
    Solvents: Methanol/Water
    Gradient: 35.0:65.0 $CH_3OH:H_2O$
       to   50.0:50.0 $CH_3OH:H_2O$
                 over 5 min.
       then   50.0:50.0 $CH_3OH:H_2O$
       to   65.0:35.0 $CH_3OH:H_2O$

    Delay Volume: 155 ul.

Solute	Obs. t_R	Calc. t_R	% diff.
Uracil	3.2	---	---
Phenol	9.0	9.3	3.3
Acetophenone	12.0	12.1	0.8
Nitrobenzene	13.0	13.0	0
Methylbenzoate	16.2	16.0	1.3

a. All retention times are in minutes
b. All solvent compositions are in volume percent.

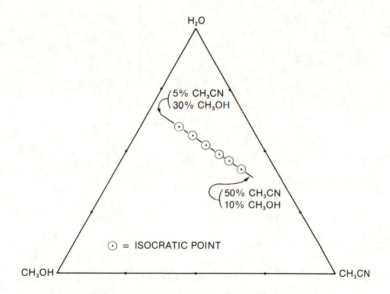

Figure 1. Ternary solvent diagram indicating mobile phase concentrations for isocratic data points. Reproduced from Ref. 22. Copyright 1985, American Chemical Society.

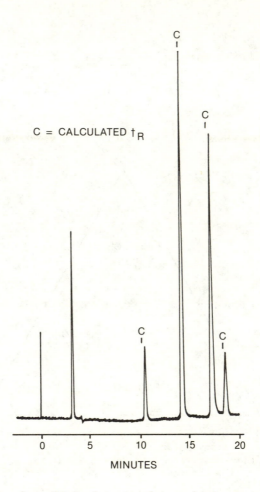

Figure 2. Microbore HPLC of test solutes showing actual and calculated (c) retention times for a 10 minute linear ternary gradient, (conditions listed in Table 4a) using a 155 μl instrumental gradient delay. Column; 250 X 1 mm; laboratory bonded Partisil 10 C_8; flow rate; 50 μl min $^{-1}$. Detector; Kratos 769, fitted with a 0.5 μl flow cell. Wavelength 254 nm, sensitivity 0.1 aufs. Pump; Brownlee MPLC micobore gradient pump. Injector; Rheodyne 7413 with a 0.5 μl injection loop. Reproduced from Ref. 22. Copyright 1985, American Chemical Society.

Table III. Ternary Isocratic Data

Point #	1	2	3	4	5	6
Uracil	3.2	3.0	3.0	3.0	2.8	2.8
Phenol	11.8	9.2	7.7	6.4	5.8	5.4
Acetophenone	21.0	15.0	11.9	9.4	8.2	7.5
Nitrobenzene	29.4	21.6	16.8	12.8	11.0	9.7
Methylbenzoate	41.0	26.6	19.3	14.0	11.8	10.4

Mobile Phase Concentration

Point#	% Acetonitrile	% Methanol
1	14.0	26.0
2	20.8	23.0
3	27.5	20.0
4	34.3	17.0
5	38.8	15.0
6	43.3	13.0

corresponding data are presented in Table IV a. The gradient used is graphically illustrated in Figure 3. Another ten minute, ternary gradient was run from 27.5% acetonitrile, 20.0% methanol to 43.3% acetonitrile, 13.0% methanol with a 155 ul instrumental delay. The results for this gradient are given in Table IV b. The results for the ternary gradients clearly show that even though ternary systems are mathematically and experimentally very complex, they can be handled with an accuracy exceeding 96%.

Quaternary Gradients. In order to test the applicability of the program and instrumentation under rigorous conditions, quaternary solvent systems were investigated for both linear and multiple linear gradient shapes. Quaternary gradients have not been exploited fully, possibly because of the tremendous complexity of such solvent systems. A solvent concentration line was again evaluated as opposed to the entire response surface. It should be noted that quaternary and higher order systems pose additional problems since their response surfaces require four or more dimensions to visualize. While a minimum of ten isocratic data points are necessary to fit the equation for aqueous solutions containing three organic modifiers, eleven data points were acquired. The high strength mobile phase reservoir contained an aqueous 12.5% tetrahydrofuran, 20.0% acetonitrile and 12.5% methanol solution and the low strength mobile phase was an aqueous 2.5% tetrahydrofuran, 2.5% acetonitrile and 25.0% methanol solution. The 11 isocratic data points used are given in Table V. Both linear and multiple-linear quaternary gradients were run. The results for a twenty minute gradient from 6.0% THF, 8.6% acetonitrile and 20.6% methanol to 11.0% THF, 17.4% acetonitrile and 14.4% methanol with a 155 ul instrumental delay are given in Table VI a. A graphical illustration for this gradient is presented in Figure 4. Changing the instrumental delay had little effect on the accuracy of the calculations.

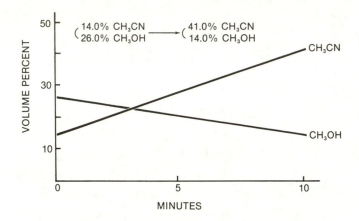

Figure 3.   Ten minute ternary gradient for the chromatogram shown
in Figure 2, as entered into gradient generator.   Reproduced from
Ref. 22.   Copyright 1985, American Chemical Society.

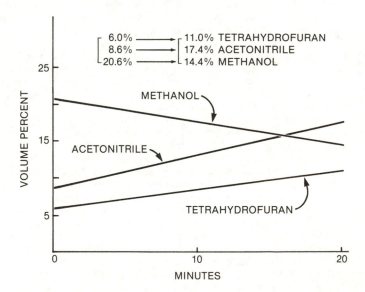

Figure 4.   Twenty minute quaternary gradient profile as entered
into the gradient generator.   Reproduced from Ref. 22.   Copyright
1985, American Chemical Society.

Table IV.   Single Segment Linear Ternary Gradient Profiled

a.

Conditions:
   Solvents: Acetonitrile/Methanol/Water
   Gradient: 14.0:26.0:60.0 $CH_3CN:CH_3OH:H_2O$
            41.0:14.0:45.0 $CH_3CN:CH_3OH:H_2O$
            over 10 min.
   Delay Volume: 155 ul.

Solute	Obs. t_R	Calc. t_R	% diff.
Uracil	3.2	---	---
Phenol	10.5	10.3	1.9
Acetophenone	14.4	14.1	2.0
Nitrobenzene	17.4	16.8	3.4
Methylbenzoate	18.8	18.3	2.7

b.

Conditions:
   Solvents: Acetonitrile/Methanol/Water
   Gradient: 28.0:20.0:52.0 $CH_3CN:CH_3OH:H_2O$
     to   43.0:13.0:44.0 $CH_3CN:CH_3OH:H_2O$
            over 10 min.
   Delay Volume: 155 ul.

Solute	Obs. t_R	Calc. t_R	% diff.
Uracil	3.0	---	---
Phenol	7.4	7.5	1.4
Acetophenone	10.7	10.8	0.9
Nitrobenzene	13.9	14.1	1.4
Methylbenzoate	15.0	14.9	0.1

    Finally, a multiple linear quaternary gradient was run consisting of a first segment from 6.0% THF, 8.6% and 20.6% methanol to 10.0% THF, 15.6% acetonitrile and 15.6% methanol in 10.0 minutes followed by 10.0 minute segment to 11.0% THF, 17.4% acetonitrile and 14.4% methanol. Like the previous quaternary gradients the THF and acetonitrile concentrations rise from the initial to the final concentrations while the methanol concentration falls. The results for the multiple linear quaternary gradient are presented in Table VI b.
    The overall accuracy of the quaternary gradient experiments was slightly worse than for the simple binary gradients, but in no case was the error between predicted and observed more than 6%. It was felt, given the complexity of the system, that these errors were sufficiently small for prediction purposes.

Multiple Column Experiments. The accuracy of the developed program was checked by performing three binary linear gradient, coupled column experiments. Prior to conducting the gradient experiments, four isocratic data points were acquired for the five test solutes. These isocratic data are presented in Table VII.
    The first two gradient coupled column experiments were performed with a C_{18} column first (injector end) connected to a phenyl phase column. The first linear binary gradient run using

Table V.   Quaternary Isocratic Data

Point #	1	2	3	4	5	6
Compound	Retention Time (minutes)					
Uracil	3.1	3.0	3.1	3.0	3.0	2.9
Phenol	17.0	15.8	14.8	13.8	13.0	12.0
Acetophenone	21.0	18.4	17.0	15.8	14.6	13.4
Nitrobenzene	35.4	32.4	30.0	27.6	25.6	23.5
Methylbenzonate	43.0	38.2	34.4	31.0	28.2	25.4

Point # (cont.)	7	8	9	10	11
Compound	Retention Time (minutes)				
Uracil	3.2	3.2	2.8	3.0	3.0
Phenol	11.7	11.2	10.0	9.4	8.8
Aectophenone	13.0	12.3	11.1	10.2	9.6
Nitrobenzene	22.3	20.7	19.0	16.6	15.6
Methylbenzoate	23.8	21.7	19.8	16.9	16.0

Mobile Phase Composition

Point #	% Tetrahydrofuran	% Acetonitrile	% Methanol
1	5.50	7.75	21.25
2	6.00	8.63	20.63
3	6.50	9.50	20.00
4	7.00	10.38	19.38
5	7.50	11.25	18.75
6	8.00	12.13	18.13
7	8.50	13.00	17.50
8	9.00	13.88	16.88
9	9.50	14.75	16.25
10	10.50	16.50	15.00
11	11.00	17.38	14.38

Table VI. Linear Quaternary Gradient Data

---

a. Single Segment Linear Profile

Conditions:
  Solvents: THF/Acetonitrile/Methanol/Water
  Gradients:
          6.0:8.6:20.6:64.8 THF:CH_3CN:CH_3OH:H_2O
  to   11.0:17.4:14.4:57.2 THF:CH_3CN:CH_3OH:H_2O
  over 20 min.

Delay Volume: 155 ul.

Solute	Obs. t_R	Calc. t_R	% diff.
Uracil	3.0	---	---
Phenol	14.2	14.2	0
Acetophenone	15.8	16.0	1.3
Nitrobenzene	24.1	23.4	2.9
Methylnrnzoate	25.4	24.8	2.4

---

b. Multiple Linear Quaternary Profile

Conditions:
  Solvents: THF/Acetonitrile/Methanol/Water
  Gradient:
          6.0:8.6:20.6:64.8 THF:CH_3CN:CH_3OH:H_2O
  to   10.0:15.6:15.6:58.8 THF:CH_3CN:CH_3OH:H_2O
  over 10 min.

Delay Volume: 155 ul.

Solute	Obs. t_R	Calc. t_R	% diff.
Uracil	3.0	---	---
Phenol	13.8	13.6	1.4
Acetophenone	15.2	14.9	2.0
Nitrobenzene	22.4	21.6	3.6
Methylbenzoate	23.5	22.8	3.0

Table VII.   Isocratic Retention Data

Column: 10 cm C_{18}

Compound	Volume Percent Methanol			
	65%	55%	45%	35%
Uracil	1.4	1.4	1.4	1.6
Guaifenesin	2.0	2.4	3.3	6.4
p-Cresol	2.5	3.8	5.9	11.2
Nitrobenzene	3.3	5.0	8.2	15.2
p-Ethylphenol	3.3	6.0	11.2	25.0
Methylbenzoate	4.1	7.2	14.0	31.6

Column: 15 cm Phenyl phase

Compound	Volume Percent Methanol			
	65%	55%	45%	35%
Uracil	2.0	2.0	2.0	2.2
Guaifenesin	3.5	5.0	7.4	15.5
p-Cresol	4.1	6.4	10.0	13.5
p-Ethylphenol	5.5	10.0	18.2	42.6
Nitrobenzene	6.6	11.0	18.2	36.0
Methylbenzoate	6.6	12.6	24.2	58.2

this configuration was a 30 minute gradient from 50% to 70% methanol. The delay volume for all gradients was held constant at 220 ul. The actual and calculated retention times under these conditions for the five test solutes are given in Table VIII a.

A second gradient was run from 45% to 70% methanol in 30 minutes with both the actual and calculated retention times for the test solutes being given in Table VIII b. Excellent agreement between the experimentally determined retention times and those calculated by the program were observed for both cases.

The order of the columns was then reversed so that the solute entered the 15 cm phenyl phase column and eluted from the 10 cm C_{18} column. A forty minute gradient from 30% to 70% methanol was run. The resulting data are presented in Table IX. Again, excellent agreement between the actual and calculated retention times occurred for all five compounds.

The average accuracy of the calculations for these coupled column experiments are actually better than those previously described for single column experiments. While this may at first be surprising it must be realized that the theory used for a single column experiment is merely a simplification of that necessary for the coupled column experiment. It is reasonable therefore to look for experimental differences between the two sets of experiments performed. The most obvious differences are:

1. differences in gradient delay
2. differences in slope of gradient
3. differences in complexity of gradient conditions studied
4. differences in the ability to fit capacity factor vs
     mobile phase conditions for C_8 and C_{18} stationary phases
5. possible instrumental differences.

Table VIII.   Actual and Calculated Retention for Coupled
C_{18} and Phenyl Columns

Conditions:
  Column Order and Length:
        1. C_{18}(10 cm)
        2. Phenyl (15 cm)
    Flow rate:
        50 ul/min
    Delay Volume:
        220 ul

a. Gradient Conditions: 50% to 70% Methanol over 30 minutes

Compound	Actual t_R	Calc. t_R	% diff.
Uracil	3.5	---	---
Guaifenesin	9.0	8.7	3.3
p-Cresol	12.2	12.1	0.8
p-Ethylphenol	17.9	18.0	0.5
Nitrobenzene	17.9	17.8	0.6
Methylbenzoate	21.0	21.0	0.0

b. Gradient Conditions: 45% to 70% Methanol over 30 minutes

Compound	Actual t_R	Calc. t_R	% diff.
Uracil	3.6	---	---
Guaifenesin	11.2	10.8	3.6
p-Cresol	15.0	14.8	1.3
Nitrobenzene	21.0	20.9	0.5
p-Ethylphenol	21.5	21.6	0.4
Methylbenzoate	24.4	24.5	0.4

Table IX.   Actual and Calculated Retention for Coupled
Phenyl and C_{18} Columns

Conditions:
        Column Order and Length:
            1. Phenyl(15 cm)
            2. C_{18}(10 cm)
        Flow Rate:
            50 ul/min
        Delay Volume:
            220 ul

Gradient Conditions: 30% to 70% Methanol over 40 minutes.

Compound	Actual t_R	Calc. t_R	% diff.
Uracil	3.7	---	---
Guaifenesin	20.4	20.5	0.5
p-Cresol	25.4	24.5	2.0
Nitrobenzene	32.2	31.9	0.9
p-Ethylphenol	33.8	33.5	0.9
Methylbenzoate	37.0	36.8	0.5

The actual determination of which factors are most important
in determining the accuracy of calculated retention will require
further investigation. The fact remains, however, that the
developed approach and corresponding computer programs can predict
retention under complicated experimental conditions with acceptable
accuracy. The program as written can accommodate up to 6 coupled
columns, each with their own particular k' vs mobile phase
composition function, over any desired gradient function. This
represents then a nearly universal approach to the calculation of
retention for solutes under all possible elution conditions.

With the program and derived mathematics, the scientist can
calculate the retention times and peak widths for any set of
solutes, run under any conditions ranging from isocratic to complex
gradients with multiple columns. Given this mathematical tool, the
problem of determining which chromatographic conditions will
achieve the desired separation, within any constraints, becomes the
next problem to be approached. The solution to such an
"optimization" problem is not easy, however, the necessary
universal mathematical tools are now available for researchers in
this area to develop approaches to optimization strategies.

Conclusions. The algorithms derived using integration by parts on
a larger computer avoid the problems of attempting to derive exact
mathematical solutions to the gradient integral equations.
Computation times are short, even for complicated gradient shapes
and long elution times. Calculations required at most only a few
seconds. In addition, excellent flexibility is achieved using this
computational approach, since virtually any k' vs solvent
composition relationship can be accommodated. The same type of
flexibility is expected to be found for different gradient shapes,
especially those of unusual profiles not readily described by any
particular function. Multiple linear gradients were utilized for
this work, since most new instruments construct gradients by
linking linear segments. However, complicated gradients of any
mathematical form can be handled as readily with little programming
effort.

The ability to calculate a solute's retention for experiments
involving both solvent and stationary phase programming should
allow for increased interest in these types of experiments.

Literature Cited

1. Haddad, P.R.; Drouen, A.C.J.; Billiet, H.A.H. and De Galan,
    L., J. Chromatogr. 1983, 282, 71.
2. Debets, H.J.G.; Bejema, B.L. and Doornbos, Anal. Chim. Acta
    1983, 151, 131.
3. Drouen, A.C.J.; Billiet, H.A.H. and de Galan, L. Anal. Chem.
    1984, 56, 971.
4. Jandera, P. and Churacek, J., J. Chromatogr. 1974, 91, 223.
5. Jandera, P. and Churacek, J., J. Chromatogr. 1974, 93, 17.
6. Jandera, P. and Churacek, J., J. Chromatogr. 1975, 104, 9.
7. Jandera, P. and Churacek, J., J. Chromatogr. 1979, 170, 1.
8. Jandera, P. and Churacek, J., J. Chromatogr. 1974, 91, 207.

9.   Jandera, P.; Churacek, J., J. and Svoboda, L., J. Chromatogr.
     1980, 192, 37.
10. Jandera, P. and Churacek, J., J. Chromatogr. 1980, 192, 19.
11. Jandera, P. and Churacek, J., J. Chromatogr. 1980, 192, 1.
12. Jandera, P.; Churacek, J. and Colin, H., J. Chromatogr. 1981
     214, 35.
13. Jandera, P. and Churacek, J., J. Chromatogr. 1975, 104, 23.
14. Schoenmakers, P.J.; Billiet, H.A.H. and De Galan, L.
     Chromatographia, 1982, 15, 205.
15. Schoenmakers, P.J.; Billiet, H.A.H.; Tijssen, R. and De Galan,
     L., J. Chromatogr. 1978, 149, 519-537.
16. Schoenmakers, P.J.; Billiet, H.A.H. and De Galan, L., J.
     Chromatogr. 1983, 282, 107.
17. Snyder, L.R. in "High Performance Liquid Chromatography,
     Advances and Perspectives", Cs. Horvath, ed., vol. 1, pp.
     207-316, Academic Press, New York, 1980
18. Borowko, M.; Jaroniec, M.; Narkeiwicz, J.; Patrykiejew, A. and
     Rudzinski, J., J. Chromatogr. 1978, 153, 309.
19. Borowko, M.; Jaroniec, M.; Narkiewicz, J. and Patrykiejew, A.,
     J. Chromatogr. 1978, 153, 321.
20. Hartwick, R.A.; Grill, C.M. and Brown, P.R., Anal. Chem. 1979,
     51, 34.
21. Snyder, L.R. and Saunders, D.L., J. Chromatogr Sci. 1969,
     7, 195.
22. Tomellini, S.A.; Hartwick, R.A. and Woodruff, H.B., Anal.
     Chem. 1985, 57, 811.
23. Meyer, R.F. and Hartwick, R.A., Anal. Chem. 1984, 56, 2211.

RECEIVED July 2, 1985

12

Manipulation of Stationary-Phase Acid–Base Properties by a Surface-Buffering Effect
Boronic Acid–Saccharide Complexation

C. H. Lochmüller and Walter B. Hill

P. M. Gross Chemical Laboratory, Duke University, Durham, NC 27706

The presence of residual amine groups in surface bound, silica-based phenylboronic acid
phases lowers the apparent pK_a of the acid
groups. This "surface buffering" effect permits
boronate-saccharide complexation chemistry to
occur at much lower pH values than is typically
the case. The broader implications of the de-
liberate use of such effects are discussed.

Lochmüller, Wilder and Marshall (1) observed an apparent
lowering of the pKa for surface-bound quinazolines using photothermal spectrometric titration approach and further evidence for
the mechanism of this change in acid base behavior has recently
been reported by Lochmüller and Hill (2). It appears that site-
site interactions (charge repulsion) and "surface buffering" by
residual amine groups combine to produce the apparent change.
Matlin and Davidson (3) also observed interactions between the
neutral species picramidopropyl and propylamine in a "mixed",
bonded phase by photothermal spectrometry.
    One of the problems with silica-based chromatographic
materials is their relatively high solubility in mobile phases of
pH higher than 7.5 units. It is interesting to speculate that at
least for some important ionic equilibria, the acid-base chemistry
of bound molecules might be manipulated by deliberately intro-
ducing a "surface buffering" effect. We chose to explore this
possibility by studying the complexation of saccharides by boronic
acid anion because of the importance of saccharides and the normally high pH conditions required (pH 8-10) Gilham had postulated
that the local positive or negative charges on cellulose influenced the pKa of phenylboronic acid at low mobile phase ionic
strength (4). In addition, Lochmüller and Amoss first demonstrated the advantage of "mixed" bonded phases in systems where
the complexation constants are large and contribute to poor transfer kinetics (5). Later Karger used this approach to improve the
performance of bound metal complex phases (6). Partial derivatization of amine silicas might be expected to improve the efficiencies obtained with boronate phases by the same reasoning.

0097–6156/86/0297–0210$06.00/0

Because of their relative abundance and their widespread importance in biology and medicine, mixtures of saccharides, nucleosides and nucleotides have been separated using high-performance liquid chromatography (HPLC) with amino, cyano, n--alkane, silica and ion-exchange columns (7-9). Mobile-phase additives have also been used to enchance the saccharide selectivity on both normal and reversed phases as well as on cation exchange resins (10-12). Phenylboronic acid-diol derivatives have also been analyzed by gas chromatography and mass spectroscopy (13) and the interaction between boronic acids and diols has been extensively studied (14). Aqueous boric acid mobile phases were initially used to separate diols by paper chromatography and by electrophoresis (15). The incorporation of phenylboronic acid as a mobile phase additive was shown to increase the Rf values for most saccharides to a greater extend than that seen with boric acid (16). Phenylboronic acid derivatives also display a large bacteriostatic effect compared to boric acid (17). After Gilham initially demonstrated the effectiveness of boronic acid-substituted cellulose in the separation of complex mixtures of nucleosides (4)other boronic acid stationary phases have been prepared using polystyrene, cellulose and silica gel (18-23). Although these stationary phases are quite selective towards diols, they require alkaline mobile phases for adequate retention which severely limits the use of silica gel matrices. The present study investigates the use of boronic acid-substituted, amine-modified, silica gel matrices for the separation of saccharides and nucleosides under neutral conditions.

Phenylboronic acid is a Lewis acid (24) whose acidity is influenced by substituents on the aromatic ring and the phenylboronate anion has a tetrahedral structure (see Figure 1). The equilibrium between the phenylboronate anion and a diol is shown in Figure 2. Phenylboronic acids also interact with amines although this interaction is quite weak (25). The various crystalline structures of phenylboronic acid and monosaccharides and nucleosides that have been postulated (26,27) demonstrate the phenylboronic acids can form complexes with a variety of diols. Although most of these crystalline phenylboronic complexes are air stable, the saccharide complexes rapidly hydrolyze in water or in alcohols and the nucleoside complexes hydrolyze in water in less than 15 minutes. Since these derivatives decompose in water, the existence of an air-stable boronate-diol complex is not a guarantee of complexation under chromatographic conditions (16). A more accurate predictor is the magnitude of the formation constant for a polyol with phenylboronic acid in water (24).

EXPERIMENTAL

Materials. The saccharides, 4-bromotoluene, magnesium turnings, n--bromosuccimide, n-propylamine and benzoyl peroxide were purchased from Aldrich Chemical Company (Milwaukee, WI, U.S.A.) and the nucleosides were purchased from Sigma Chemical Company (St. Louis, MO, U.S.A.). Boric acid and n-butanol were obtained from Mallinckrodt Chemical Company (Paris, KY, U.S.A.).

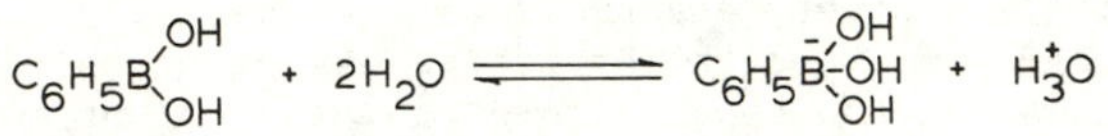

Figure 1.   The Equilibrium between Phenylboronic Acid and Water.

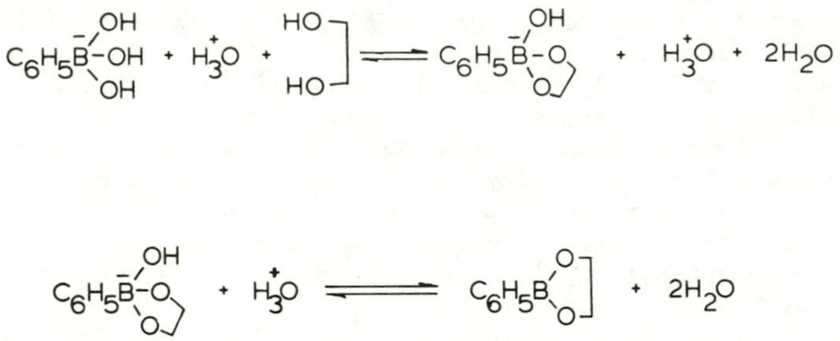

Figure 2. The Equilibrium between the Phenylboronate Anion and a Diol.

3-aminopropyltriethoxysilane (A0750) and 3-aminopropyl-
dimethylethoxysilane (A0735) were purchased from Petrarch Systems,
Inc. (Briston, PA, U.S.A.) Partisil 10 Silica Gel (s = 323 m/g;
pore diameter = 93A) was obtained from Whatman Chemical Separation, Inc. (Clifton, NJ, U.S.A.) Methanol and water were Omni-
solv–HPLC grade from MCB (Cincinnati, OH, U.S.A.), and the 0.05 $\underline{M}$
phosphate buffers were prepared with Omnisolve – HPLC grade water.
Synthesis. 4-Tolylboronic Acid was prepared from 4-bromotoluene,
magnesium turnings and tributylboronic acid according to the procedure of Bean and Johnson (28). The yield was 56% after several
recrystallizations from water. The addition of $\underline{N}$-bromosuccinimide
to 4-tolyboronic acid with dibenzoyl peroxide in anhydrous carbon
tetrachloride gave an 85% yield of 4-(ω-bromomethyl)phenylboronic
acid (29).

The synthesis of both aminated silica phases consisted of
adding either 3-aminopropyltriethoxysilane (for Phase B) or 3-
aminopropyldimethylethoxysilane (for Phase D) to dried (130°C)
Partisil 10 and refluxing for eight hours in sodium-dried toluene
with stirring over dry nitrogen gas. After Soxhlet-extraction
with methanol, the aminated silica gels were stored in a vacuum
desiccator until use. The physical characteristics of these stationary phases are listed in Table I.

Both aminated phenylboronic stationary phases (Phases B and
D) were prepared by adding a 20% excess of 4-(ω-bromomethyl)-
phenylboronic acid to the respective aminated Partisil 10 phase
and refluxing in sodium-dried toluene with pyridine (see Figure
3). After Soxhlet-extracting with methanol for forty-eight hours,
the phases were stored in a desiccator until use.

α-(3-Aminopropyldimethylethoxysilane)-4-tolyboronic acid was
prepared from an equimolar solution of 3-aminopropyldimethyleth-
oxysilane and 4-(ω-bromomethyl)phenylboronic acid at room temperature in sodium-dried toluene. After several recrystallizations
from hexane the structure was confirmed using proton nuclear magnetic resonance and infrared spectroscopy (Yield – 65%, m.p. –
172°C). Phase E was prepared by adding this silane to dried
Partisil 10 and refluxing for eight hours, the phase was stored in
a desiccator until use. The physical characteristics of all of
these stationary phases are listed in Table I.

α-($\underline{n}$-Propylamino)-tolylboronic acid (the model compound) was
synthesized by adding a 0.1 molar solution of 4-(ω-bromomethyl)-
phenylboronic acid to a 1.0 molar solution of freshly distilled
$\underline{n}$-propylamine at 50°C with stirring over dry nitrogen gas. After
distilling off the solvent and the excess reagent, the product was
recrystallized several times with isopropanol and the structure
confirmed by proton nuclear magnetic resonance and infrared spectroscopy (Yield – 72%; m.p. – 196°C). The pKa of the model compound was determined to be 10.4 according to the potentiometric
titration method of Torssell (30).

Chromatography

The chromatographic system incorporated a Varian (Walnut Creek,
CA, U.S.A.) Model 5000 liquid chromatograph, a Varian CDS 111L
data system and a Valco injection valve fitted with a 10 l. injection loop. Solute elution was monitored with a Perkin Elmer

Table I: Physical Properties of Partisil 10 Stationary Phases

Phase I.D.	A	B	C	D	E
Initial Amine Concentration moles/m^2 (x10^6)	3.27	3.27	1.48	1.48	0.815
Phenylboronic Acid Concentration moles/m^2 (x10^6)	--	0.946	--	0.475	0.815
Residual Amine Concentration moles/m^2 (x10^6)	--	2.32	--	1.00	0.00
Residual Silanol Concentration moles/m^2 (x10^6)	4.73	4.73	6.52	6.52	7.19
Ratio of Residual Amines to Phenylboronic Acid	--	2.46	--	2.11	0.00
Initial Amine % Reaction	40.90	40.90	18.50	18.50	10.20
Phenylboronic Acid % Reaction	--	28.90	--	32.10	--

*Initial concentration assumed to be 8.00 moles/m^2.

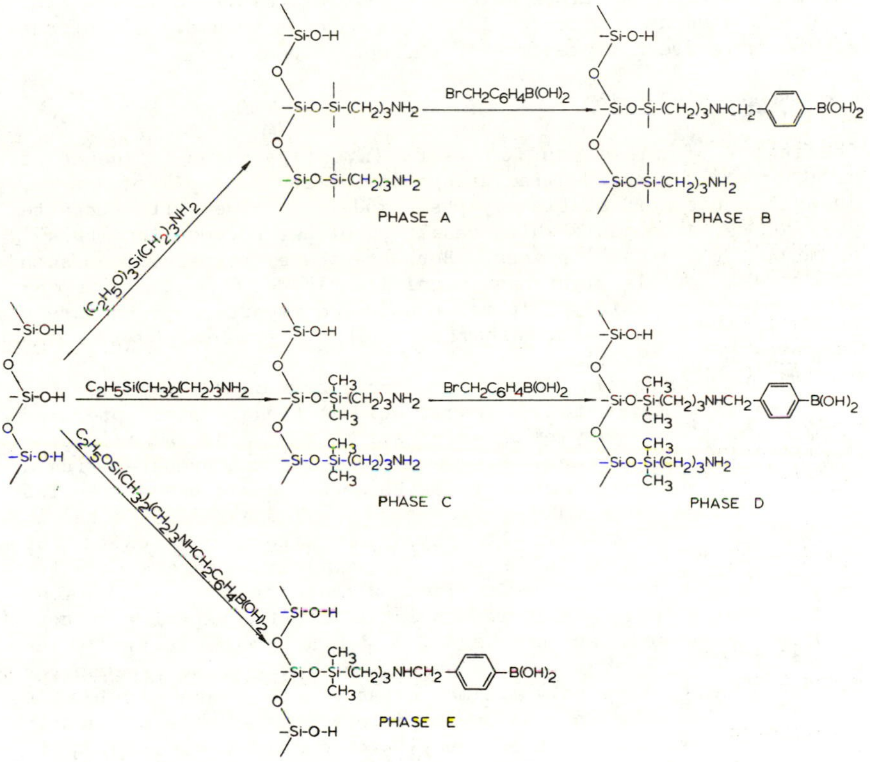

Figure 3. The Synthesis of Phyenylboronic Acid Bonded Phases on Partisil 10.

(Norwalk, Conn., U.S.A.) LC-85B UV/Vis detectro (cell volume =
1.4 µl; response time = 20 mS.) and a Laboratory Data Control
(Riviera Beach, FL, U.S.A.) RefractoMonitor Model 1107. The
stationary phases were upward slurry packed at 9000 psig. with
90:10 (v/v) methanol: water into a 25 cm. x 4.6 mm. I.D. 316
stainless steel column fitted with 2 µm frits. The average plate
number, N, for each column was calculated from the mean values of
multiple injections of mannitol, sorbitol and sucrose with a to-
tally aqueous mobile phase at 1.0 ml./min. The average N for
Phases B, D and E were 340, 1812, 1500 plates, respectively.

Chromatographic conditions were 25°C, 1.0 ml./min. and 50-100 atm.
^2H_2O was used as a measure of the column dead volume. All of the
solutes were freshly prepared in the mobile phase.

Results and Discussion

The intent of these studies is to investigate the influence of
neighboring, unreacted amine groups on the apparent pKa of bound,
phenylboronic acid stationary phase models. Amine silicas can be
prepared using reagents which result in either polymer (or "bulk")
or monomeric ("brush") phases. Boronate phases prepared from such
different materials might have significantly different local en-
vironments. In addition, it was considered important to prepare a
phase in which only phenylborate would be present by using a
phenylboronic acid silane.

Three stationary phases were prepared on Partisil 10: an
α-(n-propylamino)-4-tolylboronic acid stationary phase prepared
from a "bulk" or polymeric amine silica (Phase B), an α-(n-propylamino)-4-tolylboronic acid stationary phase prepared from a
"brush" or a monomeric amine phase (Phase D), and a phase prepared
directly using an α-(dimethylethoxypropylaminosilane)-4-tolyl-
boronic acid (Phase E). The physical characteristics of the
phases are listed in Table IV. The capacity factors (k') for
selected saccharides and nucleosides on these columns with water
as the mobile phase are listed in Table II. The capacity factors
(k') for these solutes on Phase B and D are much greater than
those on Phase E whereas the amino stationary phases themselves
(Phases A and C) have only minimal capacity for these saccharide
solutes. The order of elution for interacting solutes is identical to that observed in other phenylboronic acid chromatographic
systems (19,20) although saccharides with unfavorable diol forma-
tions and the 2'-deoxyribose derivatives have little capacity on
any of these columns. The chromatograms of a mixture of three
interacting saccharides on Phases D and E are shown in Figure IV.

Comparison of the intercolumn capacity factors is made more
meaningful by normalization for the boronic acid surface concentration (moles/m2) of the phase (see Table III). These results
indicate a large percentage increase in normalized capacity factors for Phases B and D with respect to Phase E which clearly
demonstrates the influence of residual amines on solute retention.
The residual amines (found only in Phase B and D) are promoting
phenylboronate-solute interactions by lowering the apparent pKa of
the phase either by direct interaction with boronic acid moieties
or by indirect buffering of the surface environment. (See Fig. 5.)

Table II: The Capacity Factors Of Saccharides And Nucleosides In Water

$$k', 100\% \text{ H}_2\text{O}$$

Solute	Phase A	Phase B	Phase C	Phase D	Phase E
Saccharides:					
L-(+)-Arabinose	0.01	0.42	0.02	0.42	0.08
D-Cellibiose	0.01	0.01	0.01	0.05	0.01
2-Deoxy-D-Ribose	0.02	0.15	0.02	0.16	0.10
Dulcitol	0.02	2.79	0.01	3.34	0.26
Fructose	0.05	3.07	0.03	4.14	0.26
L-(−)-Fucose	0.02	0.14	0.02	0.14	0.10
D-(+)-Galactose	0.08	0.36	0.03	0.12	0.05
D-Glucose	0.02	0.14	0.02	0.05	0.03
Inositol	0.02	0.04	0.02	0.08	0.03
Lactose	0.02	0.09	0.03	0.12	0.03
D-(+)-Maltose	0.02	0.01	0.02	0.05	0.01
Maltotriose	0.02	0.01	0.02	0.04	0.01
D-Mannitol	0.02	2.33	0.02	3.99	0.26
D-(+)-Mannose	0.02	0.22	0.02	0.21	0.05
α-D-Melibiose	0.02	0.27	0.02	0.09	0.02
D-Raffinose	0.02	0.01	0.02	0.07	0.01
D-(−)-Ribose	0.06	1.61	0.04	2.22	0.18
Sorbitol	0.04	4.14	0.05	7.15	0.45
Sucrose	0.01	0.01	0.01	0.05	0.02
D-(+)-Xylose	0.02	0.44	0.02	0.08	0.05
Nucleosides:					
Adenosine	0.19	7.89	0.20	16.59	2.05
Cytidine	0.11	4.36	0.13	13.15	1.03
2-Deoxyadenosine	0.22	0.04	0.15	0.99	0.68
2-Deoxycytidine	0.13	0.21	0.16	0.64	0.34
2-Deoxyuridine	0.03	0.09	0.08	0.32	0.09
Guanine	0.22	5.47	0.32	14.33	1.34
Thymidine	0.04	0.08	0.05	0.38	0.22
Uridine	0.05	4.15	0.06	11.86	0.84

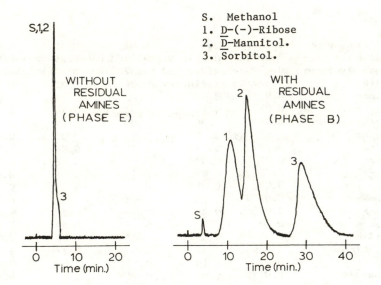

Figure 4. The Separations of a Mixture of Saccharides on
Phenylboronic Acid Stationary Phases.
Mobile Phase = water.
Flowrate = 1.0 ml./min.
Ultraviolet Detection at 195 nm.

Table III. Percentage Increase in Normalized Capacity Factors for
Strongly Interacting Saccharides and Nucleosides on Phenylboronic
Acid Stationary Phases

% Increase in Normalized k'

Solute	Phases D & B	Phases D & E	Phases B & E
Adenosine	319	1287	231
Cytidine	501	2082	263
Dulcitol	139	2072	810
Fructose	168	2653	926
Guanine	421	1736	252
D–Mannitol	241	2556	679
D–(–)–Ribose	175	2065	687
Sorbitol	244	2622	692
Uridine	469	2331	327

Table IV. Capacity Factors of Saccharides and Nucleosides on Phase D

k', T=25°C.

Solute	Water	pH 6.0	pH 5.0	pH 4.0
Saccharides:				
L-(+)-Arabinose	0.42	0.12	0.05	0.04
D-Cellibiose	0.05	0.06	0.00	0.00
2-Deoxy-D-Ribose	0.16	0.07	0.01	0.01
Dulcitol	3.34	0.66	0.49	0.15
Fructose	4.14	0.50	0.15	0.13
L-(-)-Fucose	0.14	0.11	0.07	0.06
D-(+)-Galactose	0.12	0.10	0.04	0.03
D-Glucose	0.05	0.06	0.01	0.03
Inositol	0.08	0.07	0.01	0.02
Lactose	0.12	0.08	0.06	0.04
D-(+)-Maltose	0.05	0.04	0.00	0.01
Maltotriose	0.04	0.05	0.00	0.00
D-Mannitol	3.99	1.07	0.41	0.14
D-(+)-Mannose	0.21	0.11	0.06	0.03
α-D-Melibiose	0.09	0.08	0.02	0.02
D-Raffinose	0.07	0.07	0.00	0.01
D-(-)-Ribose	2.22	0.41	0.15	0.12
Sorbitol	7.15	2.13	0.78	0.25
Sucrose	0.05	0.09	0.01	0.02
D-(+)-Xylose	0.08	0.08	0.02	0.05
Nucleosides:				
Adenosine	16.59	3.65	2.23	0.91
Cytidine	13.15	2.55	0.95	0.34
2'-Deoxyadenosine	0.99	0.89	0.81	0.52
2'-Deoxycytidine	0.64	0.29	0.23	0.12
2'Deoxyuridine	0.32	0.22	0.15	0.10
Guanine	14.33	2.81	2.08	0.58
Thymidine	0.38	0.25	0.20	0.15
Uridine	11.86	1.69	1.67	0.31

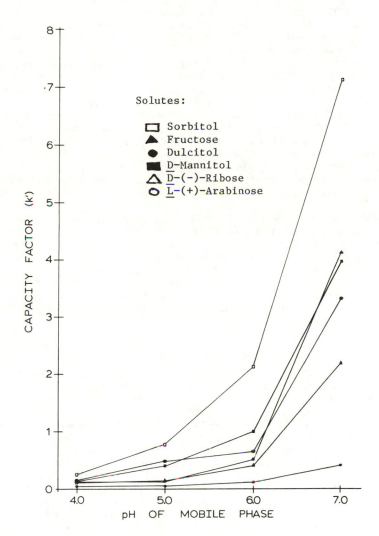

Figure 5.   The Dependence of Saccharide Capacity on a Phenylboronic Acid Stationary Phase (Phase D) on Mobile-Phase pH.
Flowrate = 1.0 ml./min.
Refractive Index Detection.

Even though the residual amine: phenylboronic acid ratio (from Table I) is higher for Phase B than for Phase D, the average normalized k's for interacting solutes on Phase D are larger than on Phase B. There are several possible explanations for this observed dependence of the capacity on the silane backbone struc- ture: 1. The polymeric backbone (Phases A and B) is composed of some amine moieties which remain inaccessible both to reaction with α-bromo-4-tolyl-boronic acid and to solvent modifications. Therefore, the ratio is an inflated estimate of the actual amine moieties available for interaction with the other moieties or with solvent. 2. The polymeric backbone could consist of terminal amines having more freedom of movement than in the monomeric phase. These surface-bound moieties would therefore be expected to be more accessible to solvent modification which would weaken the interactions between neighboring bound molecules. 3. Surface polymerization could also increase the average pore diameter of the bounded phase (by plugging smaller pores) reducing the number of site-site interactions and thus raising the pKa of the phase (2).

The relatively slow kinetics of complexation has been pro- posed as a cause of the generally low efficiencies of boronate stationary phase systems (19). The plate numbers of interacting solutes in these studies are about 90% of otherwise observed and assymetry is unchanged. The latter is unusual since previous work showed tailing ascribed to the presence of a "non-linear isotherm" (20). Perhaps the combination of surface buffering and mixed bonded phase effect creates a more favorable, more uniform sorp- tion environment.

CONCLUSION

The observed differences in normalized diol capacity factors in- dicate that residual surface amines do lower the apparent pKa of phenylboronic stationary phases by either site-site interactions or by the buffering of the surface environment. Incorporating the buffer into the surface structure instead of in the mobile phase is advantageous in lowering the mobile-phase pH required to achieve retention on weakly basic stationary phases, for collec- ting fractions in preparative chromatography and in extending the lifetime of chromatographic equipment (especially columns with silica substrates). Although most of the solute capacity in a chromatographic system depends on the mobile-phase composition and the major stationary-phase moiety, some capacity may result from residual surface species. Even if these residual functions do not interact directly with the solutes, they can modify the surface environment resulting in a change in the overall performance of the stationary phase. Similar surface buffering effects should be of use in other types of ionic equilibria where it is desirable to alter the apparent ionization constant. A more complete under- standing of these stationary-phase surface interactions should lead to the further refinement of separation strategies.

Acknowledgments

This work was supported, in part, by a grant (to CHL) from the National Science Foundation, CHE-85-000658.

Literature Cited

(1) C.H. Lochmüller, S.F. Marshall and R.W. Wilder. Photoacoustic Spectroscopy of Chemically Bonded Chromatographic Stationary Phases, Anal. Chem., **52**: 19–23 (1980).

(2) C.H. Lochmüller and W.B. Thill, Jr. Dependence of site-site interactions on silica pore diameter in amine-modified stationary phases. Anal. Chim. Acta. **157**: 65–71 (1984).

(3) R.S. Davidson, W.J. Lough, S.A. Matlin and C.L. Morrison. Photo-acoustic spectroscopic evidence for site-site interactions in a bifunctional surface-bonded phase. J. Chem. Soc. Chem. Comm. **11**: 517–518 (1981).

(4) H.L. Weith, J.L. Wiebers and P.T. Gilham. Synthesis of cellulose derivatives containing the dihydroxyboryl group and a study of their capacity to form specific complexes with sugars and nucleic acid compounds. Biochem. **9**: 4396–4401 (1970).

(5) C.H. Lochmüller and C.W. Amoss. 3-(2,4,5,7-Tetranitro-fluorenimina)-propyldiethoxysiloxane--A highly selective, bonded -complexing phase for high-pressure liquid chromatography. J. Chrom. **108**: 85 (1975).

(6) B. Feibush, M.J. Cohen, and B.L. Karger. The role of bonded phase composition on the ligand-exchange chromatography of dansyl-D,L-amino acids. J. Chrom. **282**: 3–26 (1983).

(7) S.R. Abbott. Practical aspects of normal-phase chromatography. J. Chrom. Sci. **18**: 540–550 (1980).

(8) M. Ryba and J. Beranek. Liquid chromatographic seperations of purines, purimidines and nucleosides on silica gel columns. J. Chrom. **211**: 337–346 (1981).

(9) M.W. Taylor, H.V. Hershey, R.A. Levine, K. Coy and S. Olivelle. Improved method of resolving nucleotides by reversed-phase high-performance liquid chromatography. J. Chrom. **219**: 133–139 (1981).

(10) C.H. Lochmüller and W.B. Hill, Jr. Saccharide separations in reversed-phase high-performance liquid chromatography using n-alkyl amine nobile-phase additives. J. Chrom. **264**: 215–222 (1983).

(11) K. Aitzetmüller. Applications of an HPLC amine modifier for sugar analysis in food chemistry. Chromatographia **13**: 432–436 (1980).

(12) L.A.Th. Verhaar and B.F.M. Kuster. Improved column
     efficiency in chromatographic analysis of sugars on
     cation-exchange resins by use of water-triethylamine eluents.
     J. Chrom. **210**: 279-290 (1981).

(13) C.F. Poole, S. Singhawaugcha and A. Zlatkis. Substituted
     benzeneboronic acids for the gas chromatographic
     determination of bifunctional compounds with electron-capture
     detection. J. Chrom. **158**: 33-41 (1978).

(14) K. Torssell. The chemistry of boronic and borinic acid.
     Progr. Boron Chem. **1**: 369-415 (1964).

(15) G.R. Barker and D.C. Smith. Paper chromatography of some
     carbohydrates and related compounds in the presence of boric
     acid. Chem. Ind. 19-20 (1954).

(16) E.J. Bourne, E.M. Lees and H. Weigel. Paper chromatography
     of carbohydrates and related compounds in the presence of
     benzeneboronic acid. J. Chrom. **11**: 253-257 (1963).

(17) W. Seaman and J.R. Johnson. Derivatives of phenylboric acid,
     their preparation and action upon bacteria. J. Am. Chem.
     Soc. **53**: 711-723 (1931).

(18) V.K. Akparov and V.M. Stepanov. Phenylboronic acid as a
     ligand for biospecific chromatography of serine proteinases.
     J. Chrom. **155**: 329-336 (1978).

(19) M. Glad, S. Ohlson, L. Hansson, M. Mansson and K. Mosbach.
     High performance liquid affinity chromatography of
     nucleosides, nucleotides and carbohydrates with boronic
     acid-substituted microparticulate silica. J. Chrom. **200**:
     254-260 (1980).

(20) S.A. Barker, B.W. Hatt, P.J. Somers and R.R. Woodbury. The
     use of poly(4-vinylbenzeneboronic acid) resins in the
     fractionation and interconversion of carbohydrates. Carbo.
     Res. **26**: 55-64 (1973).

(21) R.R. Maestas, J.R. Prieto, G.D. Kuehn and J.H. Hageman.
     Polyacryamide-boronate beads saturated with biomolecules: A
     new general support for affinity chromatography of enzymes.
     J. Chrom. **189**: 225-231 (1980).

(22) C.A. Elliger, B.G. Chan and W.L. Stanley. p-Vinylbenzene-
     boronic acid polymers for separation of vicinal diols. J.
     Chrom. **104**: 57-61 (1975).

(23) V. Bouriotis, I.J. Galpin and P.D.G. Dean. Applications of
     immobilized phenylboronic acids as supports for
     group-specific ligands in the affinity chromatography of
     enzymes. J. Chrom. **210**: 267-278 (1981).

(24) J.P. Lorand and J.O. Edwards. Polyol complexes and structure
     of the benzeneboronate ion. J. Org. Chem. **24**: 769–774
     (1959).

(25) K. Torssell, J.H. McClendon and G.F. Somers. Chemistry of
     arylboric acids VIII: The relationship between
     physico-chemical properties and activity in plants. Acta
     Chem. Scand. **12**: 1373–1385 (1958).

(26) R.J. Ferrier. The interaction of phenylboronic acid with
     hexosides. J. Chem. Soc. 2325–2330 (1961).

(27) A.M. Yurkevich, I.Y. Kolodkina, L.S. Varshavskaya, V.I.
     Borodulina-Shvetz, I.P. Rudakova and N.A. Preobrazhenski.
     The reactions of phenylboronic acid with nucleosides and
     mononucleotides. Tetra. **25**: 477–484 (1969).

(28) F.R. Bean and J.R. Johnson. Derivatives of phenylboric acid,
     their preparation and action upon bacteria. II.
     Hydroxyphenylboric acids. J. Am. Chem. Soc. **54**: 4415–4425
     (1932).

(29) K. Torssell. Zur Kenntnis der Arylborsauren. III.
     Bromierung der Tolylborsauren nach Wohl-Ziegler. Ark. Kemi
     10: 507–511 (1956).

(30) K. Torssell. Zur Kenntnis der Arylborsauren. VII.
     Komplexbildung zwischen Phenylborsaure und Fructose. Ark.
     Kemi **10**: 541–547 (1957).

RECEIVED November 1, 1985

13

Cyclodextrin Mobile-Phase and Stationary-Phase Liquid Chromatography

L. J. Cline Love and Manop Arunyanart

Department of Chemistry, Seton Hall University, South Orange, NJ 07079

A three-phase equilibrium model which predicts capacity factor, k', with changes in cyclodextrin (CD) mobile phase concentration was found to accurately describe the retention behavior of solutes. Graphs of $1/k'$ vs. [CD] are linear, but not parallel, with intersection points about which retention order of binary mixtures reverse. CD-solute formation constants, K_f, calculated from these graphs show that the larger the K_f value, the greater the rate of change of k' with [CD], and the CD concentration at the intersection points varies depending on the amount of added methanol. The retention order using CD mobile phases changes for different types of stationary phases (CN or C-18), but the K_f values are invariant with column type. Separations using cyclodextrin stationary phases and conventional reversed phase solvents were improved by addition of buffer when separating phenolic compounds. Separation order using a conventional reversed phase C-18 column and a beta-CD stationary phase column changes because of the additional interaction involving CD-solute complexation. Dependence of k' on methanol content of the mobile phase was found to be exponential for beta and gamma cyclodextrin columns, exactly analogous to reversed phase LC on C-18 columns. Separation of enantiomers of leucine and norlecine is demonstrated using a β-CD column and a buffered aqueous methanol mobile phase.

Aqueous and methanolic solutions of cyclodextrins have been employed as mobile phases in high performance liquid chromatography (HPLC), and as stationary phases by bonding the cyclodextrin to silica packing by several workers (1-8). They have been shown to be especially suitable for separation of structural isomers, cis-trans geometric isomers, and enantiomers. Cyclodextrins (CD) are toroidal-shaped, cyclic, oligosaccharides made up of α-1,4 linked,

0097-6156/86/0297-0226$06.00/0

D-glucopyranose units. Because of the size constraints of their hydrophobic cavities, cyclodextrins act as "hosts" to form stable inclusion complexes with a variety of "guest" species. The stability of the complex is generally related to the guest's ability to physically fit within the CD cavity, and the tightness of fit is sensitive to the isomeric characteristics of the guest molecules, thus facilitating separation of very similar species. However, many other factors, such as Van der Waal forces, dipole-dipole interactions, hydrogen bonding, and hydrophobic interactions also play roles in determining the ease of complex formation ($\underline{4}$). In addition, for efficient enantioselectivity, chiral solutes must contain a substituent which can hydrogen bond with the 2-hydroxyl groups at the entrance of the CD cavity ($\underline{9, 10}$).

In conventional reversed phase HPLC, differences in the physicochemical interactions of the eluate with the mobile phase and the stationary phase determine their partition coefficients and, hence, their capacity factor, k'. In reversed-phase systems containing cyclodextrins in the mobile phase, eluates may form complexes based not only on hydrophobicity but on size as well, making these systems more complex. If 1:1 stoichiometry is involved, the primary association equilibrium, generally recognized to be of considerable importance in micellar chromatography, can be applied ($\underline{11-13}$). The formation constant, K_f, of the inclusion complex is defined as the ratio of the entrance and exit rate constants between the solute and the cyclodextrin. Addition of organic modifiers, such as methanol, into the cyclodextrin aqueous mobile phase should alter the kinetic and thermodynamic characteristics of the system. This would alter the K_f values by modifying the entrance and exit rate constants which determine the quality of the separation.

It is well known that k' decreases as the concentration of organic modifier is increased in HPLC, and it is of interest to determine if this relationship holds in CD-HPLC. Normally, k' decreases exponentially with the volume fraction, δ, of organic modifier, and a linear relationship between $\ln k'$ and [organic modifier] is described in Equation 1 below ($\underline{14}$),

$$\ln k' = \ln k^0 - S \delta \qquad (1)$$

where k^0 represents the k' value for a compound using an aqueous mobile phase. The quantity, S, depends on both the solvent strength and on the specific interactions between the solute, stationary phase and the mobile phase. Equation 1 has been shown to hold for many substances ($\underline{14-18}$). However, Hennion ($\underline{17}$) and Schoenmakers ($\underline{18}$) carried out detailed studies of the variation of k' with the volume fraction of organic solvent in water, and determined that the dependence becomes non-linear when the composition range of the mobile phase is extended. In addition, the rate of change of $\ln k'$ with volume fraction of organic solvent is dependent on the actual modifier employed. In the present study, only methanol gave a satisfactory linear relationship between $\ln k'$ with volume fraction of methanol in the mobile phase.

The use of CD mobile phases suffers the disadvantages of low solubility in water (0.016 M), even less solubility in methanolic solutions, and cost. However, cyclodextrins chemically-bonded to silica stationary phases can be used to solve many separation problems (1, 19). Several investigators have shown that optical, geometrical, and structural isomers can be separated using CD-bonded stationary phases (20-24). They concluded that CD-bonded stationary phases retained solutes via inclusion complexation and enantioselectivity, which was determined based on a three-point attachment model.

This paper describes the elution behavior of benzene, ortho-, meta-, and para-nitrophenol, naphthalene, and biphenyl in aqueous and methanolic cyclodextrin mobile phases. The inclusion complex formation constants of these selected compounds in β-CD mobile phases containing 0%, 10%, and 20% methanol are reported. The formation constants obtained using a CN column/CD mobile phase were determined and used to predict the elution behavior of the test compounds on C-18 columns. Also, the elution behavior of solutes on C-18, β-CD, and Y-CD stationary phases are compared.

Experimental

Apparatus. The HPLC system consisted of a Technicon FAST.LC high pressure pump (Technicon, Tarrytown, NJ), Model 7120 sample injector with a 20 μL injection loop (Reodyne, Inc., Cotali, CA). The HPLC analytical columns were 5 μm Supelcosil C-18 and 5 μm Supelcosil LC-CN (both 25 cm x 4.6 mm i.d. from Supelco, Inc., Bellfonte, PA), and Cyclobond I (beta-CD) and Cyclobond II (gamma-CD) (25 cm x 4.6 mm i.d., Advanced Separation Technologies, Inc., Whippany, NJ). A precolumn (12.5 cm x 4.6 mm i.d.) packed with silica gel (25-40 μm) was located between the pump and sample injector in order to saturate the mobile phase with silica to minimize dissolution of the analytical column packing. All chromatograms were recorded on a Recordall Model 5000 strip chart recorder (Fisher Scientific, Springfield, NJ).

Reagents. Certified A.C.S. spectroanalyzed methanol, phenol, m-nitrophenol, benzene, naphthalene, sodium acetate, and acetic acid were obtained from Fisher Scientific and were used as received. The β-CD, dansyl-DL-leucine, dansyl-DL-norleucine (all from Signal Chemical Co., St. Louis, MO), o-chlorophenol, m-chlorophenol, p-chlorophenol, o-nitrophenol (all from Eastman Kodak Co., Rochester, NY), p-nitrophenol, and biphenyl (from MCB, South Plainfield, NJ) were all used as received.

Procedure. Two separate procedures were employed, referred to as system I and system II below. System I consisted of a nonpolar C-18 column and a polar CN column, both used with a mobile phase containing β-CD. The mobile phase was prepared by dissolving the appropriate quantity of β-CD in distilled water, aqueous 10% methanol, or aqueous 20% methanol, followed by filtration through a 0.45 μm Nylon-66 membrane filter (Rainin Instruments, Woburn, MA). System II consisted of a nonpolar C-18, a β-CD, and a Y-CD column, and the mobile phase was water containing an organic modifier. The mobile phase was prepared by adding the appropriate quantity of methanol to distilled water containing 0.05 M sodium acetate buffer, followed by filtration through a 0.45 μm Nylon-66 membrane filter. For both systems, methanolic stock solutions were diluted to the desired concentrations with either distilled water or methanol. A mobile phase flow rate of 1.0 mL/min was used, and retention times measured from the point of injection to the peak maximum on the chromatogram. The void volumes used in the k' calculations were 2.87 mL, 2.85 mL, and 2.75 mL for the Supelcosil C-18, the Supelcosil LC-CN, and both the Cyclobond I and Cyclobond II columns, respectively.

Results and Discussion

Cyclodextrin Mobile Phases

A three-phase equilibrium model for partitioning of solute, E, between a bulk aqueous phase and dissolved cyclodextrin, and between the bulk aqueous phase and the stationary phase, L_s, allows one to derive equations relating capacity factor, k', to the molar concentration of CD. The equation given below is similar to that derived for micellar chromatography which also assumes a three-phase model (12, 13),

$$k' = \frac{\phi [L_s] K_1}{1 + K_f [CD]} \tag{2}$$

where ϕ is the phase ratio of the stationary phase volume, v_s, to that of the mobile phase, v_m, K_1 is the solute-stationary phase equilibrium constant, and K_f is the solute-cyclodextrin formation constant. This equation predicts parabolic dependence of k' on the concentration of CD, and linear dependence of 1/k' on [CD], and its validity was tested in the present study.

Dependence of retention on mobile phase composition.
The retention data for benzene, o-, m-, and p-nitrophenol, naphthalene, and biphenyl chromatographed on a Supelcosil LC-CN column using mobile phases consisting of aqueous β-CD, β-CD in 10% methanol, and β-CD in 20% methanol are shown in Table I, and are graphically shown as 1/k' vs. [CD] in Figure 1. As predicted by Equation 2, the plots are linear, and the linear regression analyses are given in Table II. The values of K_f, also given in Table II, were calculated for each solute using the slope/intercept ratios from the graphs in Figure 1. It is apparent from the graphs that the linear dependence of the solutes on mobile phase composition are not parallel, but intersect one another. This allows rapid determination of the optimum mobile phase concentration for separation of mixtures by measurement of the k' at two concentration of β-CD for each solute. The optimum concentration is determined by drawing the lines and selecting the point of maximum separation between the k' values. In addition, shorter analysis time is gained (without changing the retention behavior) by addition of an organic modifier such as methanol. One drawback on the use of added methanol is the decrease in the solubility of β-CD in methanolic solutions.

Changes in the retention order of the ortho-, meta-, and para-

Table I. Variation of Capacity Factor of Compounds on Supelcosil-CN
Columns versus Aqueous [β-Cyclodextrin] with 0, 10, and 20% Methanol

Compound	[β-Cyclodextrin] (M)				
	0	0.001	0.002	0.004	0.006
	Water				
Benzene	2.92	2.61	2.43	2.12	1.88
Phenol	2.45	2.20	2.13	1.88	1.69
o-Nitrophenol	5.24	4.85	4.54	3.91	3.50
m-Nitrophenol	6.33	5.51	4.90	4.02	3.50
p-Nitrophenol	6.24	5.18	4.49	3.53	3.21
Naphthalene	29.46	19.53	15.08	10.60	8.06
Biphenyl	74.00	21.00	12.74	7.68	5.61
	10% Methanol				
Benzene	2.07	1.76	1.57	1.28	1.08
Phenol	1.63	1.47	1.38	1.20	1.07
o-Nitrophenol	2.98	2.76	2.58	2.28	2.03
m-Nitrophenol	4.10	3.56	3.19	2.65	2.26
p-Nitrophenol	4.04	3.39	2.89	2.29	1.90
Naphthalene	14.48	9.61	7.35	4.98	3.75
Biphenyl	31.79	10.31	6.28	3.54	2.51
	20% Methanol				
Benzene	1.47	1.31	1.16	0.95	0.81
Phenol	1.06	1.02	0.96	0.88	0.81
o-Nitrophenol	1.84	1.31	1.16	0.95	0.81
m-Nitrophenol	2.49	2.37	2.19	1.94	1.73
p-Nitrophenol	2.44	2.23	2.00	1.73	1.52
Naphthalene	6.72	5.56	4.59	3.45	2.74
Biphenyl	12.73	6.68	4.50	2.76	2.01

nitrophenol isomers illustrate some of the unusual interactions occurring in the separation. In the absence of β-CD in the mobile phase, the elution order based on the measured k' values is ortho < para < meta. When β-CD is added to the mobile phase, solute molecules are included into the CD cavity, and the retention order may change depending on the strength of interaction between the CD and the solute. However, many other factors can play a role in determining the order of retention, such as β-CD concentration and the amount of added organic modifier. The changes in the order of retention of the nitrophenol structural isomers are shown in Figure 1. At a given methanol concentration, the retention order depends on the concentration of β-CD, and reversal of the retention order occurs on either side of the intersection point of the lines. The β-CD concentration at the intersection points changes depending on the amount of added methanol. These data suggest that the β-CD is forming inclusion complexes with the isomers, and that the strength of these interaction is affected by the composition of the bulk solvent system.

Retention order also can change depending on the type of stationary phase employed. Figure 2A shows the chromatographic separation pattern for several solutes using a Supelcosil LC-CN column and a mobile phase containing 0.001 M β-CD in 20% methanol aqueous solution. Under these conditions, the compounds are retained in the order phenol < benzene < ortho < para < meta-nitrophenol, and base line separation is not achieved. However, by using a Supelcosil C-18 column with all other conditions the same, base line separations are achieved (Figure 2B), but with considerably longer retention times and a different retention order (phenol < para < meta < ortho < benzene) compared to those using the CN column. This results because more polar solutes have smaller k' values on nonpolar stationary phases. Thus, one would expect benzene to be retained longer on the C-18 column than on the CN column.

Effect of mobile phase composition on K_f. The formation constant for complex formation of solute with the CD is given by the ratio of the entrance to exit rate constants of the solute with the CD cavity. This constant should be invariant with cyclodextrin concentration, but would be expected to change upon addition of organic modifiers to the CD mobile phase. Experimentally, the K_f values increased somewhat upon addition of 10% methanol, but then decreased significantly upon addition of 20% methanol (Table II). This suggests that small amounts of methanol (up to 10%) enhances the entrance rate constant over the exit rate constant, most probably because the highly interactive water molecules normally present in "empty" CD cavities are displaced by methanol molecules which would facilitate entrance of the solute. Further increases in the methanol content of the bulk aqueous phase result in a less polar bulk phase in which the solute is more soluble, thus enhancing its exit rate constant, decreasing the entrance rate constant, and decreasing the stability of the inclusion complex (smaller K_f). Generally, the magnitude of the K_f values reflects both the strength of interaction between the CD and analyte, and the rate of change of k' with [CD]. Thus, the larger the K_f value, the greater the

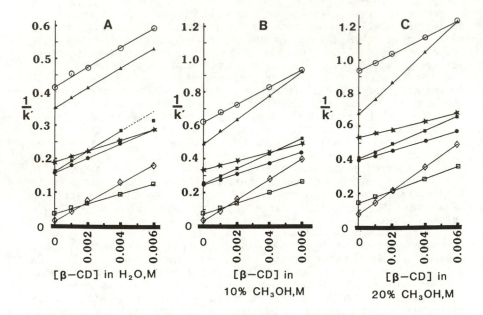

Figure 1. Dependence of 1/k' on (A) [β-CD] in water, (B) [β-CD] in 10% methanol, and (C) [β-CD] in 20% methanol, for (○) phenol, (▲) benzene, (■) p-nitrophenol, (✱) o-nitrophenol, (●) m-nitrophenol, (◇) biphenyl, and (□) naphthalene; column, Supelcosil LC-CN (25 cm x 4.6 mm i.d.); flow rate, 1 mL/min.

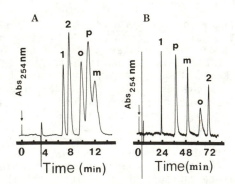

Figure 2. Characteristic separation pattern of (1) phenol, (2) benzene, (o) o-nitrophenol, (p) p-nitrophenol, (m) m-nitrophenol; column (A) Supelcosil LC-CN (225 cm x 4.6 mm i.d.), (B) Supelcosil C-18 (25 cm x 4.6 mm i.d.; mobile phase, 0.001 M β-CD in aqueous 20% methanol; flow rate, 1 mL/min; detector range, (A) 0.04 and (B) 0.02 a.u.f.s.

Table II.   Calculated Formation Constants and Statistical Analysis
for Solutes with 0, 10 and 20% Methanol/β-Cyclodextrin Mobile Phases

Compound	Slope	Intercept	% RSD		K_f (L/mol)
			Slope	Intercept	
Water					
Benzene	31.0	0.348	2.83	0.85	89
Phenol	29.6	0.414	5.25	1.26	72
o-Nitrophenol	16.0	0.190	2.58	0.81	84
m-Nitrophenol	21.4	0.160	3.54	1.77	134
p-Nitrophenol	25.7	0.167	1.33	6.50	154
Naphthalene	14.8	0.0354	1.79	2.52	418
Biphenyl	27.1	0.0190	3.85	18.93	1,432
10% Methanol					
Benzene	73.0	0.489	1.45	0.73	149
Phenol	52.8	0.620	2.32	0.68	85
o-Nitrophenol	26.0	0.336	0.71	0.19	77
m-Nitrophenol	32.8	0.246	1.26	0.57	133
p-Nitrophenol	46.5	0.250	1.25	0.78	186
Naphthalene	31.8	0.0701	0.61	0.97	468
Biphenyl	61.1	0.0349	1.19	7.07	1,751
20% Methanol					
Benzene	93.3	0.676	1.05	0.49	138
Phenol	49.2	0.939	2.41	0.43	52
o-Nitrophenol	23.3	0.539	3.07	0.45	43
m-Nitrophenol	29.9	0.397	2.82	0.72	75
p-Nitrophenol	41.5	0.411	2.22	0.76	101
Naphthalene	36.3	0.146	1.38	1.16	249
Biphenyl	69.9	0.0803	0.74	2.16	870

strength of interaction and the greater the rate of decrease in k' with increasing cyclodextrin concentration. Cyclodextrin mobile phase chromatography provides an effective way to determine inclusion complex formation constants, and their magnitudes are useful in evaluating solute-CD interactions. Also, the K_f values obtained chromatographically can be confirmed spectroscopically if identical experimental conditions are employed.

Dependence of complex formation constants on column type. According to Equation 2, one would expect to obtain identical K_f values using either a CN or a C-18 column. The data in Figure 2 revealed large differences in k' between different column types, and that lengthy experiments would be required to calculate K_f using the C-18 column. However, the K_f values obtained using the CN column should allow prediction of k' values at any concentration of CD for the C-18 column. In order to test this hypothesis, the C-18 column was first calibrated by measuring the k' values of the compounds shown in Figure 2 using a 0.004 M β-CD/20% methanol mobile phase. The k' value at zero β-CD concentration (equal to the $\phi[L_s]K_1$ intercept term) was then calculated using the measured k', the β-CD concentration, and the K_f value previously determined on the CN column. The observed k's and calculated $\phi[L_s]K_1$ terms are given in Table III. The intercept and formation constants should accurately predict k' values at any concentration of k' if Equation 2 adequately describes the system.

Table III. Test of Invariance of Mobile Phase CD-Solute Formation Constants on Different Reversed Phase Columns by Prediction of k' on a C-18 Column using K_f found on a CN Column

Compound	Experimental[a] k'	Intercept $\phi[L_s]K_1$	Predicted[b] k'	Observed[c] k'	Δ %
Phenol	4.69	5.67	5.14	5.28	2.65
o-Nitro	13.69	16.04	14.77	15.07	1.99
m-Nitro	9.92	12.92	11.23	11.67	3.77
p-Nitro	7.03	9.87	8.21	8.47	3.07
Benzene	14.10	21.88	17.15	17.36	1.20

[a] Using a C-18 column and a 0.004 M β-CD/20% methanol mobile phase; intercept calculated from experimental k', [CD], and K_f.
[b] Predicted k' for a 0.002 M β-CD/20% methanol mobile phase calculated using the intercept and K_f values.
[c] Experimentally observed k' at same mobile phase concentration.

To test the validity of Equation 2 and the constancy of the K_f values between different columns, the measured k' values from the chromatogram shown in Figure 3 were compared to the ones calculated as described above for the same mobile phase composition (Table III). The deviations between the calculated capacity factor and the observed capacity factor ranged from 1.2 to 3.8%. This good agreement strongly suggests that Equation 2 accurately describes the separation process, and that the K_f values are not affected by the nature of the stationary phase.

Cyclodextrin Stationary Phases

The role of the mobile phase composition. The ability of ionizable acids and bases to form cyclodextrin inclusion complexes or to interact with nonpolar stationary phases can be affected by altering the pH of the mobile phase (25). Equation 2 has been expanded to predict pH effects on k' using micellar mobile phases, and the same expanded equations should equally well apply to CD-mobile phase chromatography (26). Buffers added to the mobile phase not only maintain a constant pH, but can also improve efficiency, enhance resolution, improve asymmetric peaks and minimize other undesirable phenomena encountered using CD stationary phases (27). Figures 4 and 5 illustrate these effects for the separation of o-, m-, and p-nitrophenol using a gamma-CD stationary phase column with a conventional 1/4 methanol/water mobile phase, both buffered and unbuffered. The chromatogram in Figure 4 exhibits poor efficiency and lack of resolution of the ortho and meta isomers. Considerable improvement is obtained by adding 0.05 M acetate buffer, pH 4.6, to the same system (Figure 5). Note that the retention times using the buffered system are approximately half those of the unbuffered system. The three isomers can not be adequately resolved, probably because of less efficient mass transfer between the mobile and the stationary phase, and because the solutes' size are small compared to the gamma-CD cavity size, reducing the strength of the CD complex. Buffer was used in the mobile phase for all of the studies on methanol effects described below.

The three structural isomers of nitrophenol, the three isomers of chlorophenol, and the parent phenol are separated using nonpolar C-18 and beta-CD columns shown in Figures 6 and 7, respectively. Generally, the elution order using cyclodextrin stationary phases is based on the ability of eluates to form CD inclusion complexes, and the larger the solute-CD K_f, the larger the k' will be. However, the nature of the spacer arm attaching the CD to the stationary phase support, and the mobile phase composition can play important roles in the separation process. For example, the inclusion formation constants for cresol isomers have been reported in the order ortho < meta < para, and the order of retention was ortho < meta < para (1, 23). However, another literature report gave a retention order for the same cresol isomers as meta < ortho < para (20). Because CD columns have additional interactions with solutes other than strictly hydrophobic (as with C-18 columns), the elution order on the two columns would not be expected to always be the same. Inspection of the orders in Figures 6 and 7 reveal dissimilar retention orders for the isomers studied.

The effect of the amount of organic modifier in the mobile phase on the chromatographic behavior of nine test compounds was studied using three different stationary phases, namely, C-18, beta-

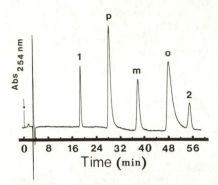

Figure 3. Characteristic separation pattern of the same compounds in Figure 2; column, Supelcosil C-18 (25 cm x 4.6 mm i.d.); mobile phase 0.002 M β-CD in 20% aqueous methanol; flow rate, 1.0 ml/min; detector range, 0.04 a.u.f.s.

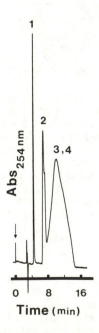

Figure 4. Characteristic separation pattern of (1) phenol, (2) p-nitrophenol, (3) m-nitrophenol, (4) o-nitrophenol; column, Cyclobond II (gamma-CD) (25 cm x 4.6 mm i.d.); mobile phase, 1/4 methanol-water; flow rate, 1.0 mL/min; detector range, 0.04 a.u.f.s.

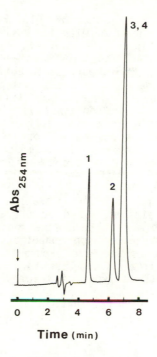

Figure 5.   Same as in Figure 4 except the mobile phase   had 0.05 M acetate buffer added (pH 4.6).

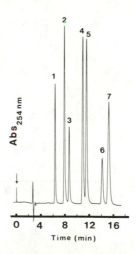

Figure 6.   Separation pattern   of (1) phenol, (2) p-nitrophenol,   (3) m-nitrophenol,   (4) o-nitrophenol, (5) o-chlorophenol,   (6) p-chlorophenol,   and (7) m-chlorophenol; column,   Supelcosil   C-18 (25 cm x 4.6 mm i.d.); mobile phase, 1/1 methanol-water with 0.05 acetate buffer (pH 4.6); flow rate, 1.0 mL/min; detector range, 0.08 a.u.f.s.

cyclodextrin, and gamma-cyclodextrin. The results listed in Table IV show a decrease in retention with increasing methanol content for all solutes. The elution order of ortho and meta-nitrophenol obtained for both the C-18 and beta-CD columns reverses by changing the mobile phase from buffered aqueous to buffered aqueous methanol. For the beta-CD stationary phase, the retention order of nitrophenol isomers is ortho < meta < para, which follows the order of the inclusion complex formation constants (Table II). It should be noted that this retention order is not the same as that previously reported in the literature of meta < ortho < para (20, 23). The lack of agreement for these isomers is most likely due to the fact that different experimental conditions were employed in the studies, and the columns likely had different β-CD loadings.

Typically plots of ln k' versus methanol content of the mobile phase is shown in Figure 8. The plots are linear, and exhibit a high degree of correlation based on the respective regression analysis given in Table V. These data indicate that the capacity factor of these test solutes on nonpolar C-18 and cyclodextrin-bonded stationary phases decrease exponentially with the methanol content of the mobile phase. Therefore, the elution behavior obtained with these three types of columns can be explained in a similar manner, and the CD columns behave in the same fashion as conventional reversed phase columns.

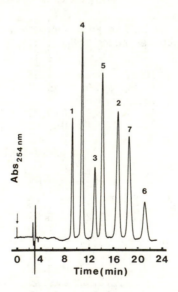

Figure 7.   Same solutes as in Figure 6; column, Cyclobond I (β-CD) (25 cm x 4.6 mm i.d.); mobile phase, 1/4 methanol-water with 0.05 M acetate buffer (pH 4.6); flow rate, 1.0 mL/min; detector range, 0.04 a.u.f.s.

Table IV. Capacity Factors for Solutes on C-18, Beta-CD and Gamma-CD
Columns Versus Methanol Content in the Mobile Phase

	% Methanol (v/v)[a]					
	0	10	20	30	40	50
Compound	k' on C-18 Column					
Phenol	21.57	10.18	6.25	3.25	2.14	1.15
o-Nitrophenol	51.47	33.74	17.81	9.19	5.45	2.90
m-Nitrophenol	69.52	25.23	13.51	6.70	3.87	1.97
p-Nitrophenol	58.16	21.11	11.53	5.69	3.29	1.69
o-Chlorophenol	92.66	39.39	20.99	10.11	5.45	2.67
m-Chlorophenol	115.02	57.89	32.51	16.07	8.52	4.08
p-Chlorophenol	105.34	51.93	29.43	14.50	7.64	3.73
	k' on Beta-Cyclodextrin Column					
Phenol	4.32	3.22	2.53	1.81	1.14	0.76
o-Nitrophenol	8.76	4.72	3.25	2.11	1.31	0.88
m-Nitrophenol	7.51	5.47	3.97	2.56	1.48	0.91
p-Nitrophenol	9.04	7.09	5.31	3.47	2.05	1.41
o-Chlorophenol	10.13	6.55	4.53	2.72	1.56	0.88
m-Chlorophenol	12.06	8.61	6.17	3.74	2.15	1.13
p-Chlorophenol	13.29	9.85	7.11	4.33	2.54	1.35
Naphthalene	56.89	29.91	19.79	10.27	5.43	2.50
Biphenyl	-	103.73	59.65	27.43	12.82	5.08
	k' on Gamma-Cyclodextrin Column					
Phenol	1.73	1.23	0.84	0.59	0.43	0.31
o-Nitrophenol	7.12	3.39	1.90	1.12	0.76	0.42
m-Nitrophenol	2.78	1.78	1.16	0.75	0.53	0.36
p-Nitrophenol	1.75	1.29	0.94	0.67	0.49	0.35
o-Chlorophenol	5.43	3.25	1.90	1.10	0.72	0.45
m-Chlorophenol	4.96	3.14	1.89	1.10	0.73	0.44
p-Chlorophenol	3.93	2.52	1.56	0.93	0.63	0.39
Naphthalene	40.89	18.27	9.26	4.32	2.32	1.13
Biphenyl	-	35.00	13.75	5.42	2.61	1.09

[a] Methanol-water in 0.05 M acetate buffer, pH 4.6.

Table V.   Statistical Analysis of Slopes, Intercepts, RSDs, and Correlation Coefficients for Graphs of k' vs. % Methanol (v/v) for
Solutes on C-18, Beta-CD, and Gamma-CD Columns

Compound	Slope (-)	Intercept	% RSD		Correlation Coefficient (-)
			Slope(-)	Intercept	
C-18 Column					
Phenol	0.057	2.978	3.53	2.05	0.998
p-Nitrophenol	0.059	4.018	2.69	1.19	0.999
m-Nitrophenol	0.069	4.059	4.72	2.43	0.996
p-Nitrophenol	0.069	3.880	4.74	2.53	0.996
o-Chlorophenol	0.070	4.449	2.18	1.03	0.999
m-Chlorophenol	0.066	4.755	1.47	0.62	0.999
p-Chlorophenol	0.066	4.657	1.40	0.60	0.999
Beta-Cyclodextrin Column					
Phenol	0.035	1.535	5.84	4.00	0.993
o-Nitrophenol	0.045	2.091	3.34	2.18	0.998
m-Nitrophenol	0.043	2.121	5.63	3.43	0.994
p-Nitrophenol	0.038	2.316	6.35	3.19	0.992
o-Chlorophenol	0.049	2.387	4.04	2.49	0.997
m-Chlorophenol	0.047	2.624	6.75	3.67	0.991
p-Chlorophenol	0.046	2.732	7.17	3.63	0.990
Naphthalene	0.611	4.089	4.10	1.86	0.997
Biphenyl	0.076	5.515	4.87	2.22	0.996
Gamma-Cyclodextrin Column					
Phenol	0.035	5.370	1.52	2.97	0.996
o-Nitrophenol	0.055	1.835	4.60	4.16	0.996
m-Nitrophenol	0.041	0.988	2.16	2.70	0.996
p-Nitrophenol	0.032	0.571	0.77	1.32	0.999
o-Chlorophenol	0.050	1.665	2.09	1.90	0.999
m-Chlorophenol	0.049	1.607	1.54	1.41	0.999
p-Chlorophenol	0.046	1.369	1.53	1.57	0.999
Naphthalene	0.071	3.656	1.64	0.97	0.999
Biphenyl	0.086	4.362	2.49	1.63	0.999

The conventional C-18 and the CD columns do interact differently with solutes of certain classes of isomers. For example, C-18 columns cannot separate enantiomers unless special additives are introduced into the mobile phase. The cyclodextrin bonded phases, however, can easily separate enantiomeric species as illustrated in Figure 9. The D and L enantiomers of dansyl-DL-leucine and of dansyl-DL-norleucine are resolved using a beta-CD column, but attempts to separate these isomers were unsuccessful using a C-18 column. The nature of the interactions between enantiomers and the cyclodextrin cavity has been described elsewhere (20, 21).

Acknowledgment

The financial support of the National Science Foundation grant No. CHE-8216878 to LJCL is gratefully acknowledged.

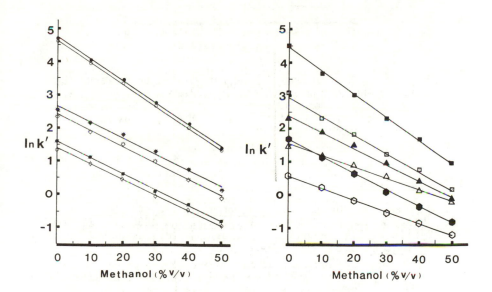

Figure 8. Dependence of ln k' on methanol content in the aqueous mobile phase of (●) m-chlorophenol, (○) p-chlorophenol, (■) o-chlorophenol, and (□) phenol on Supelcosil C-18; (⬥), (⬦), (▲), and (△) same order of compounds above on Cyclobond I (β-CD) column; (⬢), (⬡), (⬣), and (⬠) same order of compounds as above on Cyclobond II (gamma-CD) column.

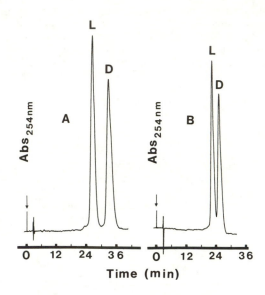

Figure 9. Separation of (A) dansyl-DL-leucine and (B) dansyl-DL-norleucine enantiomers; column, Cyclobond I (β-CD) (25 cm x 4.6 mm i.d.); mobile phase, 45/55 methanol-water with 0.05 M acetate buffer (pH 4.6; flow rate 1.0 mL/min; detector range, 0.08 a.u.f.s.

Literature Cited

1. Fujimura, K.; Ueda, T. Anal. Chem. 1983, 55, 446-50.
2. Hinze, W. L.; Armstrong, D. W. Anal. Lett. 1980, 13, 1093-2004.
3. Sybilska, D.; Debowski, J.; Jurczak, J.; Zukowski, J. J. Chromatogr. 1984, 286, 163-70.
4. Hinze, W. L. Sep. Purif. Methods 1981, 10, 159-237.
5. Sybilska, D.; Lipkowski, J.; Woycikoski, J. J. Chromatogr. 1982, 253, 95-100.
6. Uekama, K.; Hirayama, F.; Nasu, S. J. Chem. Pharm. Bull. 1973, 26 (11), 3477-84.
7. Koscielski, T.; Sybilska, D.; Jurczak, J. J. Chromatogr. 1983, 280, 131-4.
8. Nobuhara, Y.; Hirano, S.; Nakanishi, Y. J. Chromatogr. 1983, 258, 276-9.
9. Debowski, J.; Jurczak, J.; Sybilska, D. J. Chromatogr. 1983, 282, 83-8.
10. Armstrong, D. W. J. Liq. Chromatogr. 1984, 7, 353-76.
11. Armstrong, D. W.; Nome, F. Anal. Chem. 1981, 53, 1962-6.
12. Arunyanart, M.; Cline Love, L. J. Anal. Chem. 1984, 56, 1557-61.

13. Armstrong, D. W.; Stine, G. Y. J. Am. Chem. Soc. **1983**, 105, 2962-4.
14. Snyder, L. R.; Dolass, J. W.; Grant, J. R. J. Chromatogr. **1979**, 165, 3.
15. Jandera, P. J. Chromatogr. **1984**, 314, 13-36.
16. Harnisch, M.; Mockel, H. J.; Schulze, G. J. Chromatogr. **1983**, 282, 315-32.
17. Hennion, M. C.; Picard, C.; Combellas, C.; Caude, M.; Rosset, R. J. Chromatogr. **1981**, 210, 211-28.
18. Schoenmakers, P. J.; Billiet, H. A. H.; Galan, L. D. J. Chromatogr. **1979**, 185, 179-95.
19. Armstrong, D. W.; DeMond, W. J. Chromatogr. Sci. **1984**, 22, 411-5.
20. Armstrong, D. W.; DeMond, W.; Alak, A.; Hinze, W. L.; Riehl, T. E.; Bui, K. H. Anal. Chem. **1985**, 57, 234-7.
21. Hinze, W. L.; Riehl, T. E.; Armstrong, D. W.; DeMond, W.; Alak, A.; Ward, T. Anal. Chem. **1985**, 57, 237-42.
22. Tanaka, M.; Kawaguchi, Y.; Shono, T. J. Chromatogr. **1983**, 267, 285-92.
23. Kawaguchi, Y.; Tanaka, M.; Nakae, M.; Funazo, K.; Shono, T. Anal. Chem. **1983**, 55, 1852-7.
24. Tanaka, M.; Kawaguchi, Y.; Nakae, M.; Mizobuchi, Y.; Shono, T. J. Chromatogr. **1984**, 299, 341-50.
25. Tanaka, M.; Kawaguchi, Y.; Niinae, T.; Shono, T. J. Chromatogr. **1984**, 314, 193-200.
26. Arunyanart, M.; Cline Love, L. J. submitted for publication in Analytical Chemistry.
27. Melander, W. R.; Stoveken, J.; Horvath, C. J. Chromatogr. **1979**, 185, 111-27.

RECEIVED June 5, 1985

14

High-Resolution, Two-Dimensional, Gel Electrophoresis of Proteins

Basic Concepts and Recent Advances

Russell P. Tracy

Departments of Pathology and Biochemistry, University of Vermont College of Medicine, Burlington, VT 05405

Two-dimensional gel electrophoresis has become the method of choice for protein separations, when high resolution is required. Routinely, >1000 polypeptides can be separated by this technique, with >3000 possible under special conditions. This paper discusses the basic methods involved, such as gel composition and solubilization buffers, as well as recent advances, such as membrane preparations, gel scanning devices and the production of antibodies from gel spots.

Since the landmark publication of O'Farrell in 1975 (1), two-dimensional gel electrophoresis (2DGEL) has become, in terms of resolution at least, the pre-eminent protein separation technique. Usually performed as the combination of isoelectric focusing (IEF) done under denaturing conditions (9 mol/L urea, 2% non-ionic detergent) and sodium dodecyl sulfate polyacrylamide gel electrophoresis (SDS-PAGE) done in slab gels, this technique is capable of separating >1000 polypeptide gene products on a single gel, with three to four times this resolution possible, using special techniques. While the theoretical resolution of 5,000-10,000 polypeptides (1,2), is not usually achieved for technical reasons, it is none-the-less clear that 2DGEL offers the biochemist and cell biologist the highest resolution for protein separation currently available.

The historical aspects of 2DGEL have been reviewed previously (2,3), and an excellent text has recently been published, covering many aspects of 2DGEL (4). The purpose of this article is to briefly touch on basic methodologies, and review recent advances in some detail, in an

0097–6156/86/0297–0244$06.00/0

attempt to highlight those areas which have already expanded or will likely expand, the usefulness of this already powerful technique.

Background

The nomenclature "2DGEL" generally has come to mean a combination of IEF and SDS-PAGE as described above. An example of such a separation of renal cell carcinoma is shown in figure 1. Over 1000 spots are visible on this silver-stained gel. While this discussion will be limited to this particular technique, the reader should be aware that many two-dimensional electrophoretic techniques have been utilized, such as two-dimensional immunoelectrophoresis and two-dimensional SDS-PAGE (e.g.,+/- 2- mercaptoethanol).
   The advantages and disadvantages of 2DGEL are listed in Table I.

Table I. Advantages and Disadvantages of 2DGEL

Advantages	References
high resolution; >1000 components	2
sensitive	
quantity: <1 ng/spot	5
charge: <1 charge unit	6
mass: < 0.5 kd (SDS apparent MW)	6
compatible with Western blot techniques	7
compatible with global analysis methods	8
Disadvantages	
technically demanding	2
not preparative	5
reproducibility of quantitation is difficult to achieve	9

   Primary among the advantages is the excellent resolution that this technique offers. Chief among the disadvantages is the difficulty of the technique.
   The types of samples which may be analyzed by 2DGEL are varied, including virtually all animal tissues, fluids, and cells, and many microbial and plant specimens. The resulting patterns for most of these specimens are complex, and often contain more information than the researcher can immediately utilize, especially since 2DGEL is most often done to study a small subset of the total protein set present. This type of "local" analysis may be done visually, or with the aid of a small densitometer (10). However, much more complex analyses are possible ("global" analyses) with the aid of a computerized image analysis system (11). This is discussed in more detail below.

Figure 1. Silver stained two-dimensional gel of renal cell carcinoma. The acid side is oriented to the left, and large proteins are on the top. Approximately 800 spots are present.

Sample Preparation

Many different sample preparation solutions have been utilized to solubilize various specimens. We have found that most samples can be analyzed using one of the solutions listed in Table II.

Table II. 2DGEL Solubilization Solutions

Sol.No.	Sample	Components	Reference
1	serum, urine	2% SDS, 5% 2-ME, 10% glycerol, 0.05mol/L CHES, pH 9.5	12
2	lymphocytes, other cells	9 mol/L urea, 2% NP-40, 0.1 mol/L DTT, 2% 7-9 ampholytes	1,13
3	tissue, muscle	#1 above	14
4	tissue, general	9 mol/L urea, 1% SDS, 2% NP-40, 2% 2-ME, 2% 9-11 ampholytes, 2mmol/L PMSF, 2mmol/L benz-amidine, pH 9.5	15
5	RBC ghosts	4 mol/L urea, 2% NP-40, 2% 2-ME, 1% 3-10 ampholytes	16
6	plasma membranes	1% Triton-X-114 extraction; then, 9 mol/L urea, 5% 2-ME, 2% 9-11 ampholytes	17,18
7	microsomes	9.5 mol/L urea, 2% CHAPS, 0.05 mol/L DTT, 1% 3-10 ampholytes, pH 4.1	19

SDS, sodium dodecyl sulfate; 2-ME, 2-mercaptoethanol; CHES, cyclo-hexylamine sulfonic acid; NP-40, Non-Ident 40 non-ionic detergent; PMSF, phenylmethylsulfonyl fluoride; CHAPS, 3-[(3-cholamidopropyl) dimethylammonio]-1-propane sulfonate.

This list of solutions, while by no means comprehensive, should provide at least a starting point for the solubilization of virtually any animal specimen.

The most recent addition to the list is the method for membrane solubilization, reported by Willard-Gallo et al ($\underline{17}$). This preparation scheme is significantly easier to use than [^{125}I]-lactoperoxidase labelling methods, and results in better quality gels. Based upon earlier work by Bordier ($\underline{18}$), the non-ionic detergent Triton-X-114 is used to solubilize whole cells and tissues. Once solubilization is complete, the

sample is heated to 37°C, which causes the precipitation of the
Triton-X-114 at its so-called "cloud point". Any associated hydrophobic
proteins, predominantly plasma membrane proteins, are also
precipitated. This precipitate may be easily collected and analyzed by
2DGEL. The results presented by Willard-Gallo (17) clearly indicate that
this method is superior to the classical methods of [^{125}I]-labelling or
membrane isolation for many application.

Also concerning membrane proteins, Perdew et al (19) have recently
demonstrated the the usefulness of zwitterionic detergents, as
compared to non-ionic detergents, for increasing the solubility of some
membrane proteins, as well as increasing the quality of the 2DGELs
themselves.

IEF/SDS-PAGE

There are many variations of the recipes for the IEF and SDS-PAGE gels
used in 2DGEL. The standard conditions used in our laboratory are listed
in Table III; a detailed review may be found in reference 4.

Table III. Conditions for 2DGEL

Component	Specifications
IEF gels	3.5% acrylamide, 9 mol/L urea, 2% NP-40 ampholytes (~2%) :
	serum analysis: 3-10, 1.2 ml; 2-4, 0.1 ml; 9-11,0.24 ml
	tissue analysis: 3-10, 1.0 ml; 5-7, 0.25 ml
IEF electrolyte (+)	0.85% phosphoric acid
IEF electrolyte (–)	0.02 mol/L sodium hydroxide; 1.0 mol/L if a wide pH range is desired
equilibration buffer	0.01 mol/L Tris-HCl, pH 6.8, 10% glycerol, 2% SDS, trace bromophenol blue
SDS-PAGE	10-20% acrylamide, exponential gradient; 0.5 mol/L Tris-HCl, pH 8.5, 0.1% SDS
SDS-PAGE Buffer	0.02 mol/L Tris-glycine, pH8.0, 0.1% SDS

As 2DGEL has been used more and more to analyze different
specimens, researchers have developed different IEF gradients to suit
their purposes. My laboratory has reported on a broad range gradient
(4-10) for the analysis of human myeloma proteins (20), and an acidic
gradient for the non-collagenous proteins of bovine bone (21), but many

others have been reported as well. O'Farrell has reported a method for analyzing very basic proteins (which do not enter gels with the regular pH gradient), called non-equilibrium pH gradient electrophoresis (NEPHGE), a type of isotachophoresis (22). Willard et al have reported a similar system (23). In this method, samples are applied to the IEF gels from the acidic side (usually, they are applied from the basic side), and electrophoresis is started without any pre-focusing. Instead of the usual 10,000 V.hrs, the gels are processed for ~2,000 V.hrs. While not achieving equilibrium, this process does allow the researcher to visualize the basic proteins before they travel off the gel on the basic side.

Increased resolution has been the subject of many recent studies. Some of these efforts are summarized in Table IV.

Table IV. Increased Resolution in 2DGEL

Method	Reference
Multiple IEF gradients and multiple SDS-PAGE gels for the same sample	24
Both IEF and NEPHGE for the same sample	25
Larger IEF gels	26
Larger SDS-PAGE gels	27
Both larger IEF and SDS-PAGE gel	25,28
Immobilized ampholytes in IEF gel	29

Of these methods, the system of D.A.Young and co-workers offers the largest increase in resolution to date (28,30). An estimated 3,000 protein spots may be readily visualized. These "giant" gels (37x39 cm), however, are even more technically demanding than routine 2DGELs, and not readily done in large batches, as are the smaller gels (16x16 cm).

The development holding the most promise for future advances in resolution, however, involves using immobilized ampholytes in the IEF gels (29). The pH gradients prepared with these reagents are extremely stable, and can support much larger protein loads than conventional ampholytes. They are also very reproducible. To date IEF gels using immobilized ampholytes consist of relatively narrow pH ranges and are more difficult to prepare than conventional IEF gels (31). Nevertheless, the potential for this technique is impressive, and rapid advances are expected.

Visualization

Originally, the protein patterns resident in 2DGELs were visualized using autoradiography (1), or Coomassie Blue (12). Starting with the report of Merril and co-workers (32), however, staining with silver has become an important method of visualization for 2DGEL (see gel in figure 1). This method relies on the reaction of proteins in the gel with silver ions, followed by reduction to metallic silver. While the exact nature of this reaction is not known, silver staining has become popular for two reasons. First, it is ~100 times greater in sensitivity than Coomassie Blue for most proteins (33). This sensitivity is gained at some cost, however. Coomassie Blue, while relatively insensitive, has a much wider linear range (i.e., staining density per unit protein) than does silver stain, and produces images without the background staining that often occurs with silver stain.

Second, silver staining may be done in a manner which produces different colors for different proteins, apparently in a characteristic manner (34). However, the reproducibility and linearity of the color silver stain is still being debated (35). In many cases, however, the use of color silver staining may aid in the identification of a protein, or aid in differentiating two protein spots that appear close together on a gel. Recently, several commercial silver stain methods have become available, including those from Bio-Rad (Richmond, CA 94804), Helena (Beaumont TX 77704), PolyScience (Warrington PA 18976), Kodak (Rochester, NY 14650; actually a nickel stain), and Health Products, Inc. (South Haven MI 49090; a color stain).

Most workers will photograph the stained gels for later analysis. If the gels are to be analyzed by overlaying on a light box, as is commonly done, then transparencies must be made from the original negatives. (Overlaying the wet gels themselves leads to gel breakage, and limits the time of analysis, since silver gels tend to fade.) The multiple photographic steps are tedious and often lead to a loss of information. Harrison has reported the production of real-size images by direct exposure of light through the 2DGEL onto X-ray duplicating film, thereby creating a positive transparency (36). This procedure is rapid and simple, especially if automatic processing equipment is available, and yields high quality images of silver-stained gels with a minimal loss of spatial or density information. Unfortunately, Coomassie Blue stained gels may not be recorded this way due to the color sensitivity of the film.

When radioactively-labelled protein samples are available, e.g., $[^{35}S]$-

methionine-labelled cultured cells, fluorography is the most common method of visualization. Formerly, this entailed laborious washings of the gels with toxic reagents, but recently relatively simple-to-use reagents have become commercially available, which greatly simplify the procedure of impregnating the gel with the fluorophore. Two such reagents are "Enlightening", New England Nuclear (Boston MA 02118) and "Amplify", Amersham (Arlington Heights IL 60005).

Interpretation

It has been long recognized that, once the technical aspects of 2DGEL have been mastered, data analysis becomes the paramount problem (37). In a general sense, 2DGEL may be used in two ways. "Local" analysis refers to using the 2DGEL system to separate the protein components of a sample sufficiently such that quantitative or semi-quantitative information can be obtained on one or a limited number of protein spots. "Global" analysis refers to the process of obtaining quantitative information on virtually all of the protein spots present on a gel and comparing these data to similar data from other gels on a spot by spot (presumably protein by protein) basis.

  Local analysis can be performed visually on a semi-quantitative basis, or with the aid of a simple densitometer system for quantitative data. My laboratory has described a simple densitometric system based on the use of a TV camera and a microcomputer that is suitable for this type of work (10) and others have done the same (37,38), or have adopted one-dimensional scanners to this purpose (39,40).

  The major complicating factor in such analyses is the normalization of density information for quantitative purposes. Due to the inherent reproducibility problems of the technique, an "internal standard" spot, used to determine spot ratios for experimental spots, greatly facilitates the comparison of an experimental gel to a control gel. As yet, no exogenously added "internal standard" protein has been described, and various investigators have used endogenous proteins (e.g., actin, in cellular samples) as the "internal standard", the assumption being that the amount of this protein does not change between control and experimental samples. This assumption should be supported by independent measurements of the standard whenever possible. Alternatively, the absolute amount of density associated with a spot can be used, but this can increase the error associated with the measurement, especially if silver staining is used.

  Global analysis must be done with the aid of a computerized image analysis system. Several such systems have been developed by reseachers, and several others are available commercially. Figure 2 illustrates the key components of such a system, which include: 1), a scanning device for input of raw data; 2), a terminal to control the main computer (CPU) as well as a graphics computer (terminal) on the front end; 3), an option is a specialized processor, such as an array processor, to shorten computation time as the image is being manipulated; 4), a large capacity disk drive for on-line storage and a tape drive for archival storage; and, 5), a high-resolution printer/plotter for generating hard copy. Table V lists several of the systems that support global analysis of 2DGELs.

Table V. Computerized 2DGEL Image Analysis Systems Supporting Global Analysis

Research Group	System Name	Reference
Anderson et al	TYCHO	11
Garrells et al	QUEST	24
Hruschka	GELS2	41
Lemkin, Lipkin	GELLAB	42
Miller et al	LECtro	43
Ridder	–	44
Skolnick et al	–	45
Company		
Biolmage Corp.	Visage	Ann Arbor, MI
Image Analytics Corp.	LGS-30	Hockessin, DE
S.E.P. Corp. [11]	–	Genoble, France
Vickers Instruments Corp.	Scanagel	Malden, MA

[11] Societe European du Propulsion

While very sophisticated in design, none of these systems is, as yet, "fully automated" for global analysis and varying amounts of operator interaction may be required. Depending upon the system, this interaction may be needed to accomplish one or more of the following: removal of streaks and other artifacts; identification of landmark spots to facilitate image "registration" (correct alignment of the spots of one gel with the spots of another gel); and correction or removal of poorly resolved or mis-registered spots. Entering a gel may take from 30 minutes to several hours depending upon the quality of the gel, and if the image is to reside in the system as a master image or as a comparison image. Generally, the problem of internal standardization is either ignored (and absolute density values are used) or each spot is expressed as a percent of the total spot density on the gel, a value which is generally proportional to the percentage of total cell protein if several hundred of the most abundant spots are included in the analysis.

Once a system is capable of generating information for global analysis, software must be written to implement a particular strategy to analyze this information. To date, this work has lagged behind the technical develpment work, but some progress has been made. Anderson et al and Garrells have examined the differences between two cell lines by the following strategy. Each spot on each of two gels is analyzed and an integrated density value determined. The two values for each spot are then compared to each other by plotting on a graph (e.g., value for a spot on gel 1 on the x-axis and value for that spot on gel 2 on the y-axis). A statistical comparison is done as well (8,11). In preliminary investigations both groups of investigators clearly have shown that this strategy classifies cells in the order of their biological similarity to the control cells (8,11). Anderson et al then extended this approach with the use of cluster and principal component analyses (45). They have shown that, using these more sophisticated statistical methods, a panel of human cells lines are easily distinguished one from another, and that a statistical "directionality" may lead to a better understanding of co-regulated sets of proteins (45). While still in its infancy, this area of investigation holds much promise for the future analysis of 2DGEL data on a global scale.

Ancillary Techniques

Many ancillary techniques have been developed in recent years, but several stand out in their ability to extend the usefulness of 2DGEL. This final section will discuss four of these: Western blotting, antibody production from gel spots, and metabolic labelling to determine amino acid composition and tyrosine-specific phosphorylation.

Western blotting (transblotting; immunoblotting; electroblotting) refers to the technique of electrophoretically transferring proteins from gels to nitrocellulose, or other protein-binding solid support sheets, so that these proteins may be analyzed immunologically without interference from the acrylamide matrix. For a general review of this subject see reference (47). In a study designed to illustrate the general

usefulness of Western blotting with 2DGEL, Anderson et al analyzed
human plasma proteins using 30 different antisera to reveal the
locations of their respective proteins (7). While the majority of the
antisera worked well in this system, the study also pointed out that
monoclonal antibodies, relying on only a single epitope, can be
problematical if the particular epitope of interest does not survive the
2DGEL process. Many other investigators also have used Western
blotting with 2DGEL to investigate individual proteins, e.g., the
immunoblot study of low density lipoprotein receptors in individuals
with hypercholesterolemia (48).

   The production of antibodies to proteins cut from SDS-PAGE gels has
been popular for well over a decade, but only recently have investigators
attempted to produce antibodies to spots cut from 2DGELs. My
laboratory has done a study of this process and outlined methods for the
key steps (49,50). These key steps are outlined in figure 3, along with
the general research plan adopted in my laboratory. Protein spots that
are identified as interesting are cut from 2DGELs in sufficient quantity
for injection of the protein as well as [^{125}I]-labelling. If the proteins
are relatively rare in abundance, then pre-fractionation (e.g., a column
separation step), followed by a concentration step may be necessary.
Using the labelled antigen, monoclonal antibodies (MoAbs) may be
identified and used directly to study the protein of interest, or, if not
reactive with the native protein, used to isolate sufficient denatured
protein to produce a rabbit antiserum. MoAbs are preferred in this
process as a first step for several reasons: 1), a mouse requires less
antigen than does an animal such as a rabbit; 2), even using spots cut
from a 2DGEL, one may not be able to prepare absolutely pure antigen, a
problem which can be circumvented with MoAbs; 3), the quantity of
antibody which may be produced with hybridoma methods, if the correct
MoAb is found, greatly facilitates affinity isolation, assay production,
etc. This type of approach has been used by my laboratory (51), and
others (52,53), to extend the investigative power of 2DGEL.

   Latter et al have recently published the results of a study in which
cells growing in culture were labelled with 20 radioactive amino acids,
each in a different experiment, and analyzed by 2DGEL (54). The
resulting autoradiographs were studied with an image analysis system
so that amino acid compositional data could be extracted. This
information was then compared to the amino acid composition data
resident in two large data bases, using computerized search methods.
Out of the 122 spots studied in this manner, tentative identifications
could be made on 17 of them based on the amino acid composition alone.

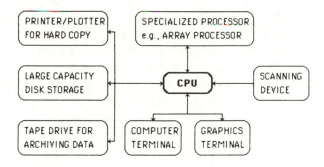

Figure 2. Essential components of a computerized system capable of supporting "global" analysis of two-dimensional gels.

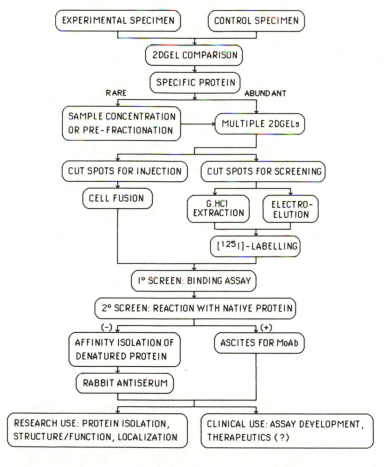

Figure 3. Essential steps in our research plan for developing monoclonal antibodies to spots cut from two-dimensional gels.

Further supporting evidence has been obtained on 7 of these proteins. The authors contend that this method is much less time-consuming than cutting out spots for amino acid analysis which, if the laboratory is set up for multiple 2DGEL analyses, is correct. However, as the authors also point out, the information that is obtained on the identity of a protein spot is tentative, and must be supported by an independent method. Latter et al used various other independent methods, such as nonactin treatment to demonstrate mitochondrial origin, co-migration with isolated proteins, and location on the gel (i.e., size and charge) to support their contentions. Others have suggested that spots may be cut fron 2DGEL for microscale amino acid sequence analysis (55), and my laboratory has in fact used this method to demonstate the identity of alpha-1-microglobulin (51). Nonetheless, the use of amino acid compositional information obtained through 2DGEL analysis may well prove to be extremely useful as a first step in identifying unknown spots present on 2DGELs.

Finally, the use of [^{32}P]-orthophosphate to label cells growing in culture has allowed the identification of many of the phophoproteins present in the cells of interest. However, recently it has become of interest to know which of these proteins are phosphorylated on a tyrosine residue, as tyrosine phosphorylation has been implicated in the regulation of cellular growth rate and in transformation. Cooper and Hunter have published a procedure for identifying which of the phosphoproteins seen on a 2DGEL is a tyrosine phosphoprotein, based on the alkali stability of the tyrosine phosphate as compared to the more common serine or threonine phosphate (56). Two-dimensional gels are run, using [^{32}P]-labelled cells as samples. After autoradiography, the gels are treated with 1 mol/L KOH for 2 hours at 55°C, neutralized, and then re-exposed. The authors report that the serine and threonine phosphorylations are much reduced when compared to the tyrosine phosphorylations which are relatively alkali stable. Thie method should prove enormously useful in identifying that subset of proteins which are phosphorylated on tyrosine residues, a subset which appears to be of regulatory significance much out of proportion to its relative abundance.

Acknowledgments

The author would like to thank Drs. N. Leigh Anderson, Jean-Paul Hoffmann, Jerry Katzmann and Donald S. Young for helpful discussions concerning this manuscript. Some of the work presented here was supported by NIH grant CA-16835J, and the Electro-Nucleonics Corp.

Literature Cited

1. O'Farrell, P.H. J.Biol.Chem. 1975, 250, 4007-21.
2. Tracy, R.P.; Anderson, N.L. In "Clinical Laboratory Annual"; Batzakis, J.; Homburger, H.A., Eds.; Appleton-Century-Crofts: New York, 1983; chapter 5.
3. Anderson, N.G.; Anderson, N.L. Clin.Chem. 1982, 28, 739-48.
4. Celis, J.E.; Bravo, R. "Two-Diemnsional Gel Electrophoresis of Proteins"; Academic: Orlando, 1984.
5. Tracy, R.P.; Katzmann, J.A.; Young, D.S. In "Electrophoresis '84"; Neuhoff, V. Ed.; Verlag Chemie: Weinheim, 1984; pp. 190-204.
6. Anderson, N.L.; Tracy, R.P.; Anderson, N.G. In "The Plasma Proteins, vol. 4"; Academic: Orlando, 1984; pp. 222-71.
7. Anderson, N.L.; Nance, S.L.; Pearson, T.W.; Anderson, N.G. Electrophoresis 1982, 3, 135-42.
8. Garrels, J.I. J.Biol.Chem. 1979, 254, 7961-77.
9. Tracy, R.P.; Currie, R.M.; Young, D.S. Clin.Chem. 1982, 28, 908-14.
10. Tracy, R.P.; Young, D.S. Clin.Chem. 1984, 30, 462-5.
11. Anderson, N.L.; Taylor, J.; Scandora, A.E.; Coulter, B.P.; Anderson, N.G. Clin.Chem. 1981, 27, 1807-20.
12. Anderson, N.L.; Anderson, N.G. P.N.A.S. (USA) 1977, 74, 5421-5.
13. Willard, K.E. Clin.Chem. 1982, 28, 1031-5.
14. Giometti, C.S.; Anderson, N.G.; Anderson, N.L. Clin.Chem. 1979, 25, 1877-84.
15. Tracy, R.P.; Young, D.S. In "Two-Dimensional Gel Electrophoresis of Proteins"; Celis, J.;Bravo, R., Eds.; Academic: Orlando, 1984; pp. 194-240.
16. Rosenblum, B.B.; Hanash, S.M.; Yew, N.; Neel, J.V. Clin.Chem. 1982, 28, 925-31.
17. Willard-Gallo, K.E.; Humblet, Y.; Symann, M. Clin.Chem. 1984, 30, 2069-77.
18. Bordier, C. J.Biol.Chem. 1981, 256, 1604-7.
19. Perdew, G.H.; Schaup, H.W.; Selivonchick, D.P. Anal.Biochem. 1983, 135, 453-9.
20. Tracy, R.P.; Currie, R.M.; Kyle, R.A.; Young, D.S. Clin.Chem. 1982, 28, 900-7.
21. Delmas, P.D.; Tracy, R.P.; Riggs, B.L.; Mann, K.G. Cacif.Tissue Int. 1984, 36, 308-16.
22. O'Farrell, P.Z.; Goodman, H.M.; O'Farrell, P.H. Cell 1977, 12, 1133-42.

23. Willard, K.E.; Giometti, C.S.; Anderson. N.L.; O'Connor, T.E.; Anderson, N.G. Anal.Biochem. 1979, 100, 289-98.
24. Garrels, J.I.; Farrar, J.T.; Burwell, C.B. In "Two-Dimensional Gel Electrophoresis of Proteins"; Celis, J.; Bravo, R., Eds.; Academic:Orlando, 1984; pp. 38-92.
25. Bravo, R. In "Two-Dimensional Gel Electrophoresis of Proteins"; Celis, J.; Bravo, R., Eds.; Academic:Orlando, 1984;; pp. 4-37.
26. Anderson, N.L., personal communication.
27. Brzeski, H.; Ege, T. Cell 1980, 22, 513-22.
28. Young, D.A. Clin.Chem. 1984, 30, 2104-8.
29. Gianazza, E.; Righetti, P.G. In "Electrophoresis '84"; Neuhoff, V., Ed.; Verlag Chemie: Weinheim, 1984; pp. 87-90.
30. Voris, B.P.; Young, D.A. Anal.Biochem. 1980, 104, 478-84.
31. Rigetti, P.G.; Gianazza, E.; Gelfi, C. In "Electrophoresis '84"; Neuhoff, V., Ed.; Verlag Chemie: Weinheim, 1984; pp. 29-48.
32. Merril, C.R.; Switzer, R.C.; Van Keuren, M.L. P.N.A.S.(USA) 1979, 76, 4335-9.
33. Merril, C.R.; Goldman, D. In "Two-Dimensional Gel Electrophoresis of Proteins"; Celis, J.; Bravo, R., Eds.; Academic:Orlando, 1984; pp. 93-111.
34. Sammons, D.W.; Adams, L.D.; Nishizawa, E.E. Electrophoresis 1981, 2, 135-41.
35. Guevara, J.; Johnston, D.A.; Ramagali, L.S.; Martin, B.A.; Capetillo, S.; Rodriguez. L.V. Electrophoresis 1982, 3, 197-205.
36. Harrison, H.H. Clin.Chem. 1984, 30, 1981-4.
37. Mariash, C.N.; Seelig, S.; Oppenheimer, J.H. Anal.Biochem 1982, 121, 388-94.
38. Garrison, J.C.; Johnson, M.L. J.Biol.Chem. 1982, 257, 13144-9.
39. Tracy, R.P.; Currie, R.M.; Young, D.S. Clin.Chem. 1982, 28, 890-9.
40. Ledvora, R.F.; Bárány, M.; Bárány, K. Clin.Chem. 1984, 30, 2063-8.
41. Hruschka, W.R. Clin.Chem. 1984, 30, 2037-9.
42. Lipkin, L.E.; Lemkin, P.F. Clin.Chem. 1980, 26, 1403-12.
43. Miller, M.J.; Olson, A.D. In "Electrophoresis '84"; Neuhoff, V., Ed.; Verlag Chemie: Weinheim, 1984; pp. 226-34.
44. Ridder, G.; VonBargen, E.; Burgard, D.; Pickrum, H.; Williams, E. Clin.Chem. 1984, 30, 1919- 24.
45. Skolnick, M.M.; Sternberg, S.R.; Neel, J.V. Clin.Chem. 1982, 28, 969-78.
46. Anderson, N.L.; Hofmann, J.-P.; Gemmell, A.; Taylor, J. Clin.Chem. 1984, 30, 2031-9.
47. Towbin, H.; Gordon, J. J.Immunol.Meth. 1984, 72, 313-40.

48. Beisiegel, U.; Schneider, W.J.; Brown, M.S.; Goldstein, J.L. J.Biol.Chem. 1982, 257, 13150-6.

49. Tracy, R.P.; Katzmann, J.A.; Kimlinger, T.K.; Hurst, G.A.; Young, D.S. J.Immunol.Meth. 1983, 65, 97-107.

50. Tracy, R.P.; Katzmann, J.A.; Young, D.S. In "Electrophoresis '84"; Neuhoff, V., Ed.; Verlag Chemie: Weinheim, 1984; pp. 190-204.

51. Tracy, R.P.; Young, D.S.; Katzmann, J.A.; Jenny, R. N.Y.Acad.Sci. 1984, 428, 144-157.

52. Celis, J.E.; Bravo, R. In "Electrophoresis '84"; Neuhoff, V., Ed.; Verlag Chemie: Weinheim, 1984; pp. 205-25.

53. Lester, E.P.; Lemkin, P.F. In "Electrophoresis '84"; Neuhoff, V., Ed.; Verlag Chemie: Weinheim, 1984; pp. 309-11.

54. Latter, G.I.; Burbeck, S.; Fleming, J.; Leavitt, J. Clin.Chem. 1984, 30, 1925-32.

55. Hunkapiller, M.W.; Strickler, J.E.; Wilson, K.J. Science 1984, 226, 304-11.

56. Cooper, J.A.; Hunter, T. Mol. Cell. Biol. 1981, 1, 165-78.

RECEIVED May 14, 1985

15

Capillary Supercritical Fluid Chromatography and Supercritical Fluid Chromatography–Mass Spectrometry

R. D. Smith, B. W. Wright, and H. R. Udseth

Chemical Methods and Kinetics Section, Pacific Northwest Laboratory, Richland, WA 99352

Capillary column supercritical fluid chromatography (SFC) and its combination with mass spectrometry (SFC-MS) constitute important new techniques for analysis of materials not amenable to GC and GC-MS. Applications include the analysis of thermally labile and high molecular weight materials. Recent developments summarized here include: the development of narrow bore (25-50 μm) fused silica capillary columns with bonded, crosslinked and deactivated stationary phases for high resolution separations; methods using rapid pressure programming for high speed capillary SFC separations; and the capability for interfacing capillary SFC with both chemical ionization and electron impact mass spectrometry. Capillary SFC with both flame ionization and mass spectrometric detectors provide detection limits in the picogram range and significantly improved separations compared to HPLC. This review describes the physicochemical basis of SFC and SFC-MS and recent applications in the analysis of fuels, high molecular weight mixtures, and labile mycotoxin and pesticide materials.

Supercritical fluid chromatography (SFC) is becoming increasingly recognized as having an important intermediate role between gas and liquid chromatography. The recent spur for the development of SFC has been the introduction of fused silica capillary columns (1,2) and the availability of commercial instrumentation. The high chromatographic efficiencies (3-5) possible with capillary columns having small diameters (< 75 μm) has attracted attention for the analysis of complex mixtures not amenable to gas chromatography (GC). These applications require both sensitive and highly selective detectors. The properties of dilute supercritical fluid solutions allow the application of both gas and condensed phase detection methods for SFC. The capillary column SFC-mass spectrometer (MS) interface, which meets the demands for both sensitivity and selectivity, has provided a further impetus for capillary SFC development (6-9).

Supercritical Fluids. In the last few years, the analytical applications of supercritical fluids have expanded rapidly, mirroring a
similar surge of interest in supercritical fluid extraction and
fractionation for various chemical engineering processes. Supercritical fluid extraction (10,11), supercritical fluid chromatography, and direct fluid injection (DFI) –mass spectrometry
(7,12,13) utilize the properties of a compound at temperatures and
pressures above its critical point. At elevated pressures this
single fluid phase has properties intermediate between those of
the gas and liquid phases and dependent upon the fluid composition,
temperature and pressure. Figure 1 gives a pressure-density relationship in terms of reduced parameters (e.g., pressure, temperature or density divided by the appropriate critical parameter),
from which only minor deviations occur for all single component
fluids. Isotherms for various reduced temperatures show the variations in density which can be expected with changes in pressure.
Thus, the density of the supercritical fluid will be typically 10^2
to 10^3 times greater than that of the gas at somewhat lower pressures. Consequently, molecular interactions increase due to the
shorter intermolecular distances. The "liquid-like" behavior of a
supercritical fluid results in greatly enhanced solubilizing capabilities compared to the "subcritical" gas, but with higher diffusion coefficients and lower viscosities compared to the corresponding liquids. Compounds of high molecular weight can be dissolved in the supercritical fluid phase at low temperatures, e.g.,
the solubility (and chromatography) of polystyrene having a molecular weight in excess of one million has been demonstrated (14).
Polymers with molecular weights in excess of several thousand are
known to be soluble in supercritical carbon dioxide which has a
critical temperature of only 31 C.
    Examples of compounds which can be used as supercritical fluid
solvents are given in Table 1. In addition to pure solvents, fluid
mixtures can be advantageously employed to increase the range of
molecular species which are soluble in a single supercritical fluid
or to allow operation in a more favorable temperature range.

TABLE 1
EXAMPLES OF SUPERCRITICAL SOLVENTS

Compound	Boiling Point (C)	Critical Temperature (C)	Critical Pressure (bar)	Critical Density (g/cm^3)
CO_2	−78.5	31.3	72.9	0.448
NH_3	−33.4	132.4	112.5	0.235
H_2O	100.0	374.2	218.3	0.315
N_2O	−88.6	36.5	71.7	0.45
Ethane	−88.6	32.3	48.1	0.203
Ethylene	−103.7	9.2	49.7	0.218
Propane	−42.1	96.7	41.9	0.217
Pentane	36.1	196.6	33.3	0.232
Benzene	80.1	288.9	48.3	0.302
Methanol	64.7	240.5	78.9	0.272
Ethanol	78.5	243.0	63.0	0.276
Isopropanol	82.5	235.3	47.0	0.273

Solubility in Supercritical Fluids. The typical relationships
observed for the solubility of a solid in a supercritical fluid as
a function of temperature and pressure is illustrated in a simplified manner in Figure 2. The solute typically exhibits a "threshold
pressure" above which solubility increases significantly (15).
The "threshold pressure" is obviously dependent upon detector
sensitivity and has little fundamental value. The region of
maximum *rate* of increase of solubility with *pressure* is typically
near the critical pressure where the rate of increase in density
with pressure is also greatest. This results from the fact that
there is often a linear relationship between log (solubility) and
density for nonvolatile compounds up to the concentrations where
solute-solute interactions become important. In contrast, where
volatility is low and at densities less than or near the critical
density, increasing temperature will typically decrease solubility.
However, "solubility" may increase at sufficiently high temperatures where the solute vapor pressure becomes significant. Thus,
while the highest supercritical fluid densities at a given pressure
are obtained near the critical temperature, greatest solubilities
within experimental pressure limitations may be obtained at somewhat lower densities but higher temperatures. On the basis of
currently available data, solubility trends in supercritical fluids
may be summarized as follows:
(1)   As with liquids, polar solutes are most soluble in polar super-
      critical fluids, although nominally nonpolar fluids can be
      remarkably good solvents for many moderately polar compounds.
      Carbon dioxide, for example, at higher pressures can exhibit
      solvating properties intermediate between pentane and methylene
      chloride.
(2)   The maximum increase in solubility with pressure usually occurs
      near the critical density (16); however, available data for
      highly polar solutes in less polar or nonpolar solvents shows
      that significantly greater densities (and pressures) may be
      necessary to obtain substantial solubility.
(3)   At normal operating pressures, half to several times the criti-
      cal pressure, solubility typically increases with pressure
      under isothermal conditions.
(4)   Under conditions of constant density, solubility generally
      increases with temperature.
(5)   A temperature increase under isobaric conditions will generally
      result in decreased solubility at pressures less than several
      times the critical pressure, and increased solubility at higher
      pressures.

Supercritical Fluid Chromatography. The origin of SFC with packed
columns goes back over two decades (17-26). The recent interest
in SFC has been due in large part to the limitations in both chromatographic efficiency and detection methods with HPLC. The introduction
of fused silica capillary columns with nonextractable stationary
phases for SFC (1) and the potential compatibility with gas phase
detection methods (13) has served to further increase the attention
given these methods.
      The advantages of SFC for improved separation and detector
interfacing accrue from the nature of the supercritical fluid (10).

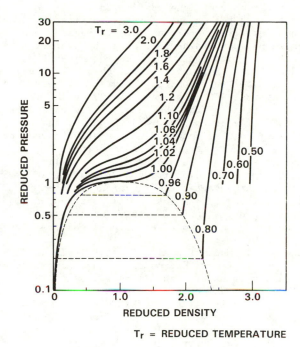

Figure 1. Typical pressure–density behavior for a pure super-critical fluid in terms of reduced parameters.

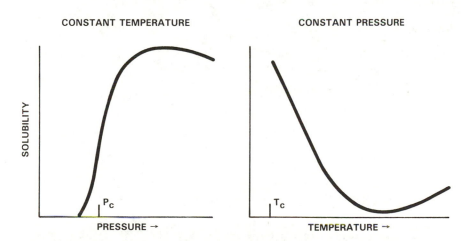

Figure 2.   Typical trends for the solubilities for solid solutes in supercritical fluids as a function of pressure or temperature (see text).

The lower viscosities and higher diffusion coefficients relative
to liquids result in the potential for significantly enhanced
chromatographic efficiency per unit time compared to HPLC. In
SFC, the mobile phase is maintained at a temperature somewhat
above its critical point (often at reduced temperatures of 1.02 to
1.4). The density of the supercritical phase is usually several
hundred times greater than that of the gas, but less than that of
the liquid at typical SFC pressures (25-500 bar). The mild thermal
conditions (determined by the choice of mobile phase) allow the
application to labile compounds. Additionally, the use of open
tubular capillary columns results in a greatly reduced pressure
drop across the column at typical linear velocities; an important
consideration if the pressure programming capability of SFC is to
be exploited to optimize selectivity. Pressure programming in SFC
provides many of the advantages of gradient elution in HPLC while
avoiding the complications impacting gradient reproducibility (27).
As is illustrated later, this also provides the basis for very
rapid pressure programming to affect high speed capillary SFC separations.

     With the development of small diameter (< 75 μm) fused silica
capillary columns coated with crosslinked and nonextractable stationary phases, high resolution separations with efficiencies approach-
ing those of conventional capillary gas chromatography have been
obtained (2,5). This is illustrated in Figure 3 which shows the
separation of the polycyclic aromatic hydrocarbon fraction of a
middle distillate fuel sample. In this separation a 20 m x 50 μm
(i.d.) fused silica capillary column coated with a bonded and
crosslinked 50% phenyl methylphenylpolysiloxane stationary phase
was used with a supercritical carbon dioxide mobile phase at 40 C.
Studies under isobaric conditions have demonstrated that more than
3,000 and 10,000 theoretical plates/m can be obtained with 50 um
and 25 um (i.d.) columns, respectively (27). Although this measure
of efficiency is inapplicable under pressure programmed conditions,
it provides the basis for the high selectivity evidenced in such
separations.

     The application of capillary SFC to such complex mixtures
clearly benefits from the use of selective detectors such as the
mass spectrometer. As is shown in the next section, the combination of SFC with mass spectrometry has many of the same advantages
of instrumental simplicity and flexibility as GC-MS.

Mass Spectrometric Detection for SFC. The advantages of coupling
a chromatographic technique with mass spectrometry are considerable,
as evidenced by the major role of GC-MS in mixture analysis. However,
in contrast to GC, where the effluent is compatible with classical
gas phase ionization methods, the ideal approach to interfacing
SFC or HPLC with mass spectrometry is not immediately obvious.
The attraction of SFC-MS or HPLC-MS combinations led to significant
interest in the late 1960's and early 1970's. The early dominance
of HPLC technology, dictated the emphasis on HPLC-MS development.
While these efforts have shown considerable success, they have not
yet yielded instrumentation comparable to GC-MS in usefulness.
The most promising current approach, utilizing "thermospray" ionization (28), is most applicable to the most polar, easily ionized

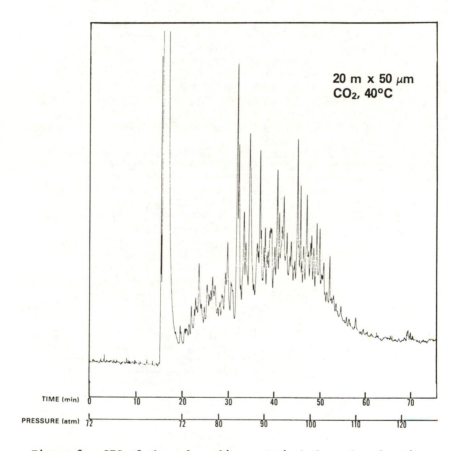

**20 m x 50 μm
CO$_2$, 40°C**

TIME (min)

PRESSURE (atm)

Figure 3.   SFC of the polycyclic aromatic hydrocarbon fraction
of a middle distillate fuel using supercritical carbon dioxide
at 40 C.

compounds. These compound classes, which generally include materials soluble in aqueous solvent systems, are also the most difficult to address by SFC. Thus, SFC-MS and thermospray LC-MS may be viewed as complimentary techniques, which together allow the analysis of virtually any compound not amenable to GC.

The development of an SFC-MS interface obviously requires that the SFC effluent undergo a pressure reduction for both ionization and subsequent mass analysis. The most direct approach to this problem involves expansion of the fluid through a nozzle and production of a collimated beam, using molecular beam skimmers, which is then ionized and analyzed using conventional electron impact ionization (EI) mass spectrometry. This approach was first proposed in the literature by Milne (29), and was a direct extension of his work with molecular beam systems, with the principle differences being that the supercritical fluid solution would have much higher densities than normal in free-jet expansions. Milne qualitatively discussed the potential problems of this approach and suggested that the high pressures could be handled with a sufficiently small diameter nozzle orifice and a multi-stage, differentially pumped vacuum system. He also indicated that the most serious potential problem of this approach would be cluster formation during adiabatic expansion. (In molecular beam studies, great care is taken to assure adiabatic, shock-free expansion conditions by using beam skimmers to sample the expanding gas into a lower pressure environment before the shock fronts form due to interaction with background gases and disrupt the process.)

Shortly after this initial report, Giddings, Myers, and Wahrhaftig presented a more quantitative evaluation of the molecular beam approach for "the gas phase isolation of nonvolatile molecules by high pressure jets for mass spectrometry" (30). These workers came to the conclusion that solvent clustering might not be significant and solute-solute association could be minimized if the expansion was sufficiently rapid. In addition, a solution of β-carotene in supercritical carbon dioxide was expanded through a small laser drilled orifice. This work provided the basis for the ambitious studies of Randall and Wahrhaftig (31-33), which provided a detailed examination of the molecular beam approach to SFC-MS interfacing.

The instrumentation constructed by Randall and Wahrhaftig utilized a three-stage vacuum system which followed the classical approach to molecular beam studies (33). This work demonstrated that solutes of low volatility could be injected into a low pressure environment. The degree of solvent clustering was much less than estimated on the basis of previous work, and found to be dependent on a number of experimental parameters. Interestingly, Randall and Wahrhaftig demonstrated that not only solute-solute clustering was negligible, as anticipated, but also that solute-solvent clustering was much less than expected. Of considerable interest was the observation that the degree of solvent-solute and solvent-solvent clustering was greatly reduced as the background pressure in the first (nozzle-skimmer) vacuum region was increased (32).

The molecular beam approach, however, does not provide a practical SFC-MS interface for routine analytical applications. The

significant complexity of this instrumentation, the multi-stage
vacuum system, and operational difficulties (such as beam alignment)
prevented serious application. More important, however, was that
this approach provided relatively poor sensitivities (μg range)
which may have been indirectly related to the frequent orifice
"plugging" problem noted by these workers (32). Our experience
has since shown that such plugging problems are often due to the
use of excessively high concentrations such that a nearly saturated
supercritical fluid solution exists which apparently leads to rapid
solute precipitation and plugging at the point of expansion.

Aside from the work of Randall and Wahrhaftig, the only other
early attempt to interface supercritical fluid chromatography was
the direct introduction work of Gouw et al (34). Although no details were provided, and additional information has not been published, this approach apparently required compound volatility for
transfer to a conventional mass spectrometer electron impact ion
source. Application appeared limited by compound volatility and
poor sensitivity.

Instrumentation for Capillary SFC-MS. The development of fused
silica capillary columns with stabilized stationary phases suitable
for SFC presented the opportunity to eliminate the disadvantages
of the molecular beam approach. The typical fluid flow rates for
50 μm i.d. columns are in the range of 1-2 1/min (as liquid);
this range can be handled by conventional two stage mass spectrometers (as typically configured with chemical ionization capability)
without the use of the liquid nitrogen cryo-pumps required for
commercial direct liquid introduction LC-MS interfaces (35).

In 1982 an effective capillary SFC-MS interface was described
(6,7). In this work supercritical fluid conditions were maintained
in the column to a region immediately adjacent to the ion source,
where the fluid was rapidly decompressed by expansion through an
orifice. The lower fluid flow rates with capillary columns avoided
the complex multi-stage pumping systems characteristic of the molecular beam approach. Rather than attempting to sample the unper-
turbed expanding gas by conventional electron impact ionization,
this direct fluid injection (DFI) method utilized the shock fronts
and rapid collisional processes in the expanding jet to disrupt
cluster species and prevent droplet formation (13). Thus, the DFI
method not only greatly simplified instrumental design but eliminated the two major drawbacks of the molecular beam approach:
clustering phenomena and inadequate sensitivity.

The DFI process requires supercritical fluid conditions to be
maintained as close to the point of ionization as possible to minimize loss of labile or nonvolatile materials. An improved understanding of the DFI process can be gained by consideration of cluster formation during expansion of a high-pressure jet through a
nozzle, as illustrated schematically in Figure 4. When an expansion occurs in a chamber with a finite background pressure (P_v),
the expanding gas will interact with the background gas producing
a shock wave system. This includes barrel and reflected shock
waves as well as a shock wave perpendicular to the jet axis (the
Mach disk). The Mach disk serves to heat and break-up the clusters
formed during the expansion process. The break-up of clusters is

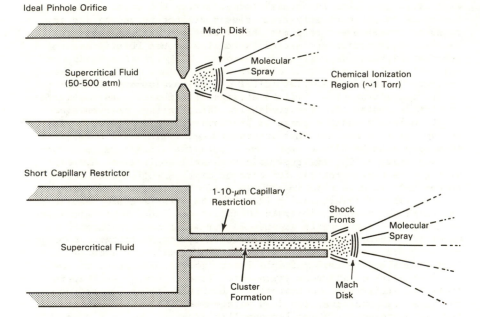

Figure 4. Schematic illustration of the Direct Fluid Injection
(DFI) process. An ideal expansion through a pinhole orifice
(top) and expansion through a capillary restriction (bottom).

further aided by the rapid collisional processes in the chemical
ionization source where the average molecule undergoes on the
order of 10^5 collisions before escaping, assuming no wall losses.
Since the solute concentration for a large peak in capillary SFC
is typically < 10 ppm, it is clear that solute clustering of neutral molecules in the ion source will usually be negligible.
(Similar arguments apply for clustering of the ionized species
since a typical average number of ion–molecule collisions in a
chemical ionization source is in the range of 10^4 - 10^6.)

The distance from the orifice to the Mach disk may be crudely
estimated from experimental work (36) as $0.67D(P_f/P_v)^{1/2}$, where D
is the orifice diameter and P_f is the fluid pressure. Thus, if
P_f = 400 bar, P_v = 1 torr, and D = 1 μm, the distance to the Mach
disk is 0.4 mm. The "droplets" formed during the expansion of a
dense gas result primarily from solvent cluster formation during
adiabatic cooling in the first stages of the expansion process.
The extent of cluster formation is related to the fluid pressure,
temperature, and the orifice dimensions. Theoretical methods for
prediction of the extent of cluster formation are still inadequate;
however, Randall and Wahrhaftig (31-33) adapted an empirical method
of "corresponding jets" (37) to attempt an estimation of the degree
of cluster formation. Unfortunately, this approach required extrapolation far beyond the range of previous data and predicted an
average cluster size several orders of magnitude larger than found
experimentally ($\sim$ 20).

In reality, the ideal pinhole type orifice shown at the top
of Figure 4 is impractical for current capillary SFC–MS interfaces.
In the initial work with 100–200 μm i.d. capillary columns this
concept was viable with 0.5 - 1.0 μm holes (6). However, even in
this case the minimum substrate thickness was $\sim$ 15 μm (and readily
deformed if not precisely supported) and had a conical channel due
to nature of the laser drilling process. Although, greater success
was achieved with orifices drilled in 50–100 μm thick material,
this typically required multiple laser "shots", was less precise
and reproducible, and more expensive. In addition, this approach
presented significant mechanical difficulties associated with producing a high pressure, low dead volume seal and the necessary
precise alignment with smaller capillaries.

In reality, all practical restrictor designs consist of short
capillary channels, schematically illustrated on the lower part of
Figure 4. A practical compromise involves the use of short lengths
(1-3 cm) of small inner diameter (5-10 μm) fused silica tubing or,
more favorably, tapered restrictors made by rapidly drawing the
terminal end of the fused silica column in a reproducible manner.
This later approach has the additional advantage of eliminating
connector dead volume. The characteristics of flow through such a
restrictor are complex and cannot be precisely predicted. However,
it is expected that most of the pressure drop occurs near the end
of the tapered region. Since initial cluster formation involves
volatile solvent molecules, heat applied in the later stage of
this expansion reduces solvent clustering and facilitates transfer
of solute molecules. However, it is clear that as the capillary
restriction becomes longer, solute clustering can become significant
with an excessively long restrictor leading to precipitation of

the solute. This can be manifested as "spikes" in the detector
output or plugging of the restrictor. While nucleation processes
will ultimately result in solute clusters, this will not be significant for sufficiently short restrictors and typical SFC concentrations. Available experimental results indicate that the DFI process
produces a gas spray incorporating the nonvolatile solute molecules under these conditions. Although experimental results for
higher temperatures and more polar fluid systems are limited, our
mass spectrometric observations show no evidence of solute cluster
formation in any systems studied to date (other than that typical
for the selected ion source pressure and temperature). Our experience with fluids as diverse as CO_2, NH_3 and supercritical water
suggest that the DFI method is viable if saturated solutions are
avoided at the point of restriction.

Current Instrumentation. The instrumentation used for SFC-MS has
been described in detail elsewhere and will be only briefly considered here (7-9, 9, 13, 28). A high pressure, programmable, syringe
pump is used to generate a pulse-free flow of a high purity fluid.
Injection utilizes an HPLC valve with volumes from 0.06 to 0.2 µL
and flow splitting at typically ambient temperatures. Since the
split occurs for a subcritical liquid, discrimination between sample
components is negligible. Split ratios range from as little as
1:3 for conventional separations on long (> 15 m) 50 µm columns to
as high as 1:80 for fast separations on 25 µm columns. The need to
minimize sample volume is clearly indicated by consideration of the
fact that the total volume of a 1 m x 25 µm column is only 0.5 µL.
     One instrument configuration utilized in this laboratory is
shown in Figure 5. In this instrument the column was mounted in a
constant temperature gas chromatograph oven, which also served to
heat the air circulated through the DFI probe. A zero dead volume
union is typically used to connect the column to a short length of
∫ 4-8 µm i.d. or contoured (tapered) fused silica restrictor. The
restrictor and probe tip are heated to compensate for cooling due
to decompression of the fluid during the DFI process.
     The mass spectrometers utilized to date in this laboratory
are conventional quadrupole mass spectrometers with two stages of
pumping. Both electron impact (EI) and chemical ionization (CI)
ion sources have been evaluated for SFC detection. The instrumentation illustrated in Figure 5 allows axial injection of the SFC
effluent, although this is not an important parameter, and different designs are used in another instrument (6-8,12,13). Methane,
isobutane or ammonia are the most frequent used CI reagent gases.
An advantage of SFC-MS with small diameter capillary columns is
that the flow rates are small enough that any CI reagent gas may
be used. Typical detection limits range from 0.1 to 10 pg depending upon the compound, analysis time, column configuration, and
CI reagent gas.
     An additional advantage of SFC-MS compared to HPLC-MS is the
recent development of a practical EI source (38). Nearly all capillary SFC-MS has utilized CI; initial attempts at EI ionization
generally resulted in greatly decreased sensitivity (7) or spectra
which contain both CI and EI components (39). However, the low
flow rates with 50 µm diameter columns, which typically correspond

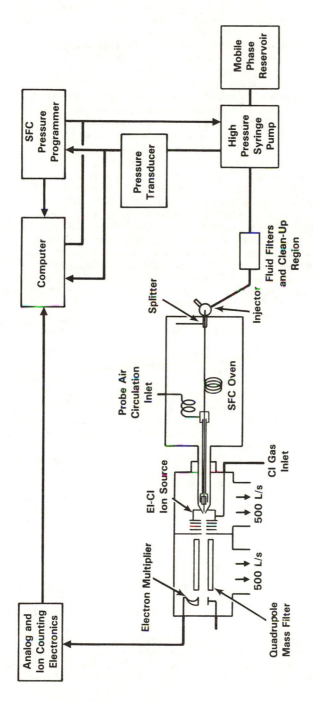

Figure 5. Schematic illustration of instrumentation for capillary SFC–MS.

to 0.1 to 0.5 mL/min of gas at standard conditions, can be made compatible with direct fluid injection using EI ionization. Most EI sources are relatively tight and are designed to operate at lower flow rates or, if a higher flow rate is used (as in GC-MS), with gases which will produce spectra quite similar to EI (i.e., charge transfer from helium). However, this "tighter" ion source can result in significant CI contributions in SFC-MS which can be reduced by a more open design (38).

A simple solution to this problem is to utilize the SFC gas flow to create a small higher pressure region adjacent to the "open" EI source (38). Thus, the fluid expands into a heated expansion region (1 cm x 0.14 cm i.d. with a 0.1 cm orifice to the ionization volume) which provides the higher pressure necessary for cluster break-up prior to the ionization region. The temperature of this region is typically 50 to 150 C higher than the mobile phase temperature (38). The expansion region efficiently directs the SFC effluent into the ionization volume of a high efficiency ionizer having an extremely open design which serves to minimize pressure in the ionization volume. The EI interface provides for efficient break-up of clusters formed during expansion and transfer of the SFC effluent to the ionization volume. While EI sensitivity is typically one to two orders of magnitude less than that obtained by CI, good mass spectra are usually obtained with < 1 ng injections and the mass spectra show no CI contributions. The flexibility in selection of the ionization method and the ability to use the existing EI spectral libraries provides an additional advantage for SFC-MS relative to LC-MS combinations employing direct liquid introduction.

Capillary Columns for SFC-MS. At present, the major limitation to broad application of capillary SFC technology is related to the availability of columns compatible with supercritical fluid mobile phases. The fused silica capillary columns used in this work were deactivated and coated with crosslinked and surface-bonded stationary phases using techniques similar to those reported by Lee and coworkers (40,41). Columns from less than 1 m to more than 20 m in length and with inner diameters of 10 to 200 μm have been examined. Column deactivation was achieved by purging with a dry nitrogen flow at 350 C for several hours followed by silylation with a polymethylhydrosiloxane. Any unreacted groups on the hydrosiloxane were capped by treatment with chlorotrimethylsilane at 250 C. After deactivation, the columns were coated with approximately a 0.15-.25 μm film of SE-54 (5% phenyl polymethylphenylsiloxane) or other polysiloxane stationary phases. The coated stationary phases were crosslinked and bonded to the deactivation layer by extensive crosslinking with azo-t-butane (41). The importance of deactivation procedures for elution of more polar compounds, such as the trichothecenes, has been demonstrated elsewhere (42).

Columns which are currently available are most amenable to less polar solute and fluid systems, and appear to be nearly ideal for fluids such as carbon dioxide. More polar fluids or fluids having higher critical temperatures appear to degrade column performance, probably due to the greater solvating power of these fluids and physical processes related to solubility in the sta-

tionary phase. Improved columns are anticipated which will be
more suitable for these demanding fluid systems.

<u>Capillary SFC-MS Development and Applications</u>. The initial applica-
tions of SFC have stressed thermally labile and higher molecular
weight mixtures not amenable to GC. Many of the early applications
of capillary SFC have utilized carbon dioxide as the mobile phase,
but the extension to alternative fluids is being actively pursued.
     In SFC both the mobile phase and the stationary phase can be
varied to enhance selectivity. The changes in selectivity are
illustrated in Figure 6 which shows SFC-MS total ion chromatograms
obtained for CO_2, C_2H_6, and N_2O mobile phase under similar
supercritical conditions. The separations were obtained at identical reduced parameters (P_r = 1.27 and T_r = 1.06). In addition to
the two n-alkanes (C_{10} and C_{15}), the separations included acetophenone (AP), N-ethylaniline (EA), naphthalene (NAP), 1-decanol
(DEC), and p-chlorophenol (CP) which were chosen to span a range
of polarities. Comparison of the chromatograms shows that significant changes in selectivity can be obtained by changing the mobile
phase.
     Capillary SFC is particularly attractive for labile compounds
which are difficult at best to analyze by GC and for which selectivities and sensitivities obtainable by HPLC are often inadequate.
Figure 7 illustrates application to labile acid and carbamate pesticides. This figure shows a capillary SFC-MS total ion chromatogram
obtained during a pressure programmed separation. Examples of
mass spectra are given in Figure 8 which compares the methane CI
(right) and ammonia CI (left) spectra for dicamba and chlorpropham
obtained during similar SFC-MS separations. The ability to select
essentially any CI reagent allows one to obtain optimum sensitivity
by minimizing fragmentation or to enhance structural information
by utilizing a more energetic CI reagent such as methane.
     An example of the capability of capillary SFC-MS for complex
mixture analysis is given in Figure 9 which shows separation of
the polycyclic aromatic hydrocarbon fraction of a diesel fuel
marine material (<u>5</u>). This separation was obtained with CO_2 at
60 C utilizing a linear pressure ramp. In general, the isobutane
chemical ionization produced predominately $(M+1)^+$ ions with smaller amounts of $(M+43)^+$ and $(M+57)^+$ ions. The solvent peak is small
since the molecular ion is not in the mass range scanned (however,
trace impurities in the carbon dioxide were detected). Typical
single ion chromatograms from this analysis are given in Figure
10. These chromatograms, show numerous isomeric species for some
components. Since the protonated molecular ions, or $(M+1)^+$ ions,
are at odd m/Z values, the molecular weights of the components are
even. Tentative identifications based on molecular weight and
chromatographic retention time for these components are indicated.
The mass spectra obtained from this analysis typically show dominant $(M+1)^+$ signals and are useful for assigning molecular weights
to the unknown components and, when used in conjunction with chromatographic retention times, serve as a basis for actual component
identifications. The material in this fraction consists primarily
of two and three ring neutral polycyclic aromatic hydrocarbons and
numerous isomers of their alkylated homologs. Various alkylated

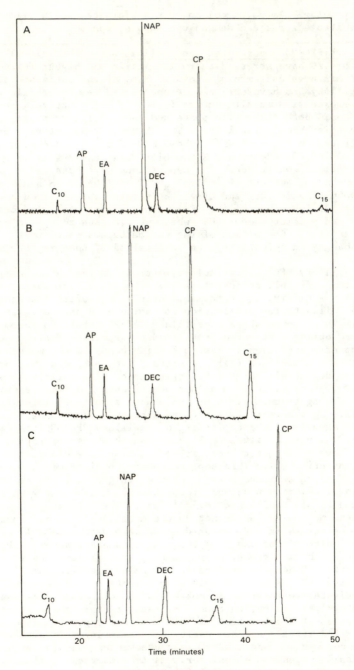

Figure 6. Comparison of capillary SFC-MS total ion
chromatograms for a mixture with (A)CO_2, (B) C_2H_6, and (C) N_2O
mobile phases.

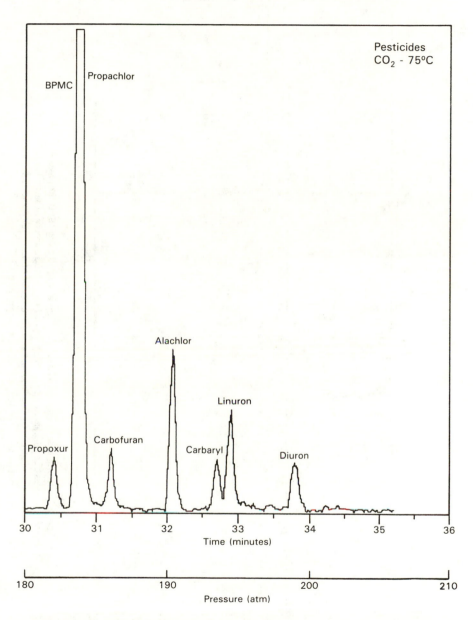

Figure 7. Capillary SFC–MS separations of an eight component pesticide mixture using a 8 m x 50 μm column at 75 C using carbon dioxide as the mobile phase.

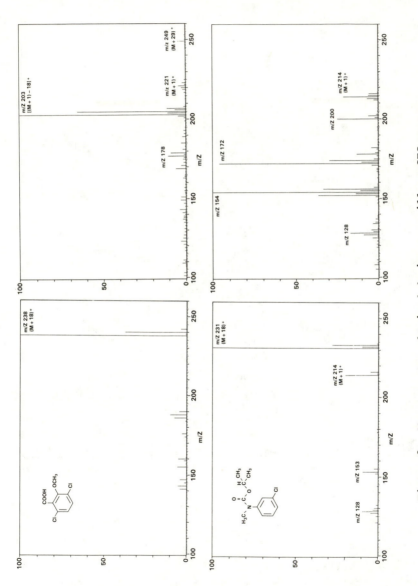

Figure 8. Mass spectra obtained during a capillary SFC separation of a pesticide mixture. Top spectra are ammonia (left) and methane (right) CI of dicamba and the bottom are ammonia (left) and methane (right) CI of chlorpropham.

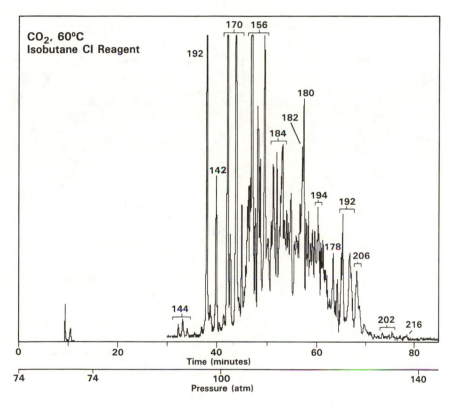

Figure 9. Total ion chromatogram obtained from the capillary
SFC–MS analysis of the polycyclic aromatic hydrocarbon fraction
of a diesel fuel marine.

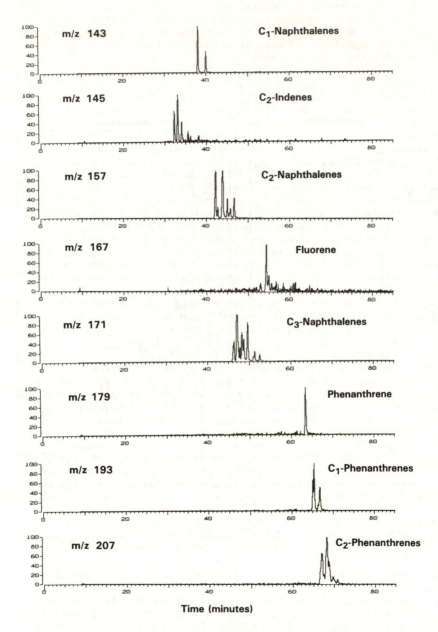

Figure 10. Typical single ion chromatograms for selected masses for the capillary SFC-MS analysis in Figure 9.

isomers of monocyclic aromatic compounds are also present (e.g.,
substituted benzenes and tetralins). Exact identifications of the
many isomers would require retention data obtained from chromato-
graphing individual standards for each component, although in some
cases more sophisticated mass spectrometric techniques can be used
to identify isomeric species (13).

High Speed Capillary SFC. Until recently the practice of SFC was
restricted to packed columns (17,20-26,43). Capillary columns are
clearly superior for most applications in gas chromatography, while
packed columns in liquid chromatography have tremendous advantages
compared to open tubular capillary columns due to the small column
diameter (< 10 μm) required to produce equivalent separations (44).
Since supercritical fluids have properties intermediate between
those of a gas and those of a liquid, one might expect that both
packed and capillary SFC methods would be competitive and the
method of choice depend on the particular separation desired.
Until recently the most common description of the state-of-the-art
for SFC would imply that packed (micro-particle) columns are best
for rapid efficient separations of less complex mixtures, whereas
small diameter (< 75 μm) capillary columns, which produce greater
numbers of total effective theoretical plates under isobaric conditions, would be preferred for complex mixture analysis. As prac-
ticed, the approximate time scale of the rapid packed column SFC
separations is on the order of two minutes (43), although high
resolution separations have required as much as 1 day. In contrast, the time scale of typical high resolution capillary column
separations is on the order of two hours (3,5,8,45,46).
     Rapid separations are clearly desirable where sufficient selec-
tivity exists for the particular component(s) of interest. The
low pressure drop across capillary SFC columns allows the solvating
properties of the fluid to be controlled by changing the pressure
on a nearly instantaneous basis (9). Additional advantages are
related to the ease of pressure control and the absence of stationary phase modification which decreases reproducibility in HPLC.
The pressure programming capability allows a significantly wider
molecular weight or compound-type range to be separated in a given
time than with packed SFC columns, which are generally restricted
to isobaric operation for rapid separations due to the large pressure drop. As with packed columns (43), capillary columns with
supercritical mobile phases can be operated at much higher than
optimum linear velocities (45,46) to obtain greater numbers of
plates/sec with only a moderate increase in plate height.
     The compatibility of capillary column SFC with mass spectrom-
etry provides the high selectivity necessary to characterize even
relatively complex mixtures in rapid separations. The average
peak width is relatively constant and is nearly ideally suited to
the maximum practical scan speeds for most quadrupole mass spectrometers ($\sim$ 0.5 sec for a reasonable mass range). Figure 11 provides a comparison between the total ion chromatograms for a high
resolution capillary SFC-MS separation and a rapid SFC-MS separation of a coal tar extract using 15m x 50 μm and 1.75 m x 50 μm
columns, respectively. The high resolution capillary SFC-MS
separation utilized supercritical carbon dioxide at 60 C and a

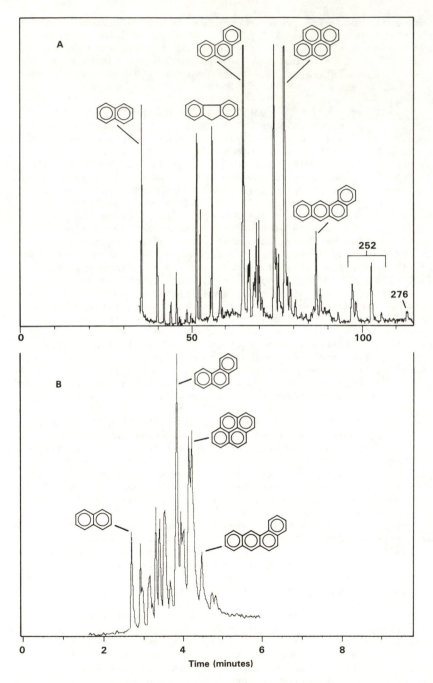

Figure 11. Comparison of a high resolution capillary SFC-MS
separation of a coal tar extract obtained using a (A) 15 m x 50
μm column with (B) a rapid separation on a 1.75 m x 50 μm
column with a 50 bar/min pressure ramp.

linear pressure ramp from 75 bar to 175 bar at 1 bar/min. The fast separation (B) utilized a linear pressure ramp of 50 bar/min starting at 75 bar and provided a level of characterization which is adequate for many purposes, especially with the inherent high selectivity of mass spectrometry. Even greater chromatographic selectivity can be obtained with smaller diameter columns and enhanced analytical selectivity can be obtained with a "soft" CI reagent gas to limit fragmentation and tandem mass spectrometric techniques (e.g., MS/MS) to resolve dissimilar compounds of the same molecular weight (47).

Although the highest chromatographic selectivity for the separation of any two compounds within a given time constraint will be obtained in an isobaric separation, the situation is quite different for multicomponent mixtures. Since an optimum pressure will exist for separation of any two components, more complex mixtures would require a different optimum pressure for each solute pair and some sacrifice of optimum selectivity. As mixture complexity increases, and the range of k' values broadens, isobaric separations become increasingly inadequate. The strength of pressure (or implicitly, density) programming is that separations can be obtained for a multicomponent mixture which will be superior to isobaric separations in terms of speed, selectivity and detector sensitivity.

Rapid pressure programmed capillary SFC techniques provide an excellent method for the fast separation of simple to moderately complex mixtures. The degree of compromise between total separation efficiency and speed of analysis can be selected and adjusted to meet the requirements for a specific separation. An example of a rapid SFC separation of a thermally labile pesticide mixture is shown in Figure 12. These carbamate and acid pesticides are generally considered to be non-gas chromatographable (50). The separation was achieved using a 1.5 m x 25 µm i.d. column, carbon dioxide mobile phase at 100 C and 135 bar initial pressure, and a pressure ramp of 50 bar/minute starting a few seconds after injection. The mobile phase linear velocity at the initial conditions was approximately 5.5 cm/sec. This separation was complete in two minutes with average peak widths of only a few seconds. By using a well deactivated column these very polar components were successfully eluted with good chromatographic peak shape using carbon dioxide as the mobile phase.

Rapid separations are enhanced by using small-diameter columns to maximize the number of theoretical plates per unit column length. This allows much shorter columns (and faster separations) to obtain the same overall efficiency. The mobile phase can also be operated at higher linear velocities in small diameter columns without significantly degrading performance. The separation efficiency (theoretical plates) for a 50 µm and 25 µm i.d. column is compared in Table II. A similar comparison using the Trennzahl (TZ) or separation number (SN) for pressure ramped analyses is described in Table III. The Trennzahl number defines the number of peaks separated by approximately twice the width at half height that can be fitted between two standards. The 25 µm i.d. column generated about four times as many theoretical plates per meter and almost ten times as many plates per minute as the 50 µm i.d. column. Although the 25 µm column was slightly shorter and the mobile phase linear velocity

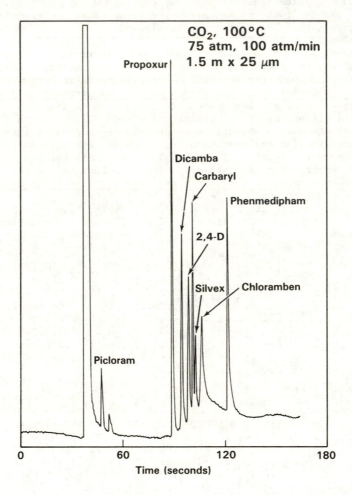

Figure 12. Fast capillary SFC separation of acid and carbamate pesticides.

Table II. Efficiency Comparison for Isobaric Separations

Column	v^a (cm/sec)	k'	$n\text{-}C_{13}$ n/m	n/min	k'	$n\text{-}C_{14}$ n/m	n/min
2 m x 50 μm	3.0	2.8	3,120	1,470	4.0	3,050	1,080
1.5 m x 25 μm	4.2	1.9	12,800	10,900	2.8	14,530	9,480

[a]Mobile phase linear velocity.

Table 3: Trennzahl Number (TZ) Comparison for Fast Pressure Ramps[a]

Column	$v(cm/sec)^b$	T_R (C_{20}) (min)	TZ ($_{26}\text{-}C_{28}$)
2 m x 50 μm	4.8	4.3	0.6
3 m x 50 μm	2.6	5.8	0.8
1.5 m x 25 μm	3.2	3.1	3.2

[a]Separations were made using a 50 bar/min pressure ramp.
[b]Mobile phase linear velocity at the initial pressure of 75 bar.

was slightly greater, the difference is striking. For pressure
ramped analyses the Trennzahl number was five times greater for
the 25 μm i.d. column and was achieved in 25% less time. A small
difference was obtained for the longer 50 μm i.d. column.

An important and unique property of capillary SFC is the capability for very fast pressure programming rates. The nearly negligible pressure drop across a short capillary column allows the fluid
density (or solvating power) to be varied on a nearly instantaneous
basis. Enhanced sensitivity results from the low effective k' at
the time of elution and the substantial "peak compression" phenomena
due to the large density increase (28).

An example of an alkane mixture chromatographed at three different pressure ramp rates is shown in Figure 13. These separations
were obtained with a 1.5 m x 25 μm i.d. column, carbon dioxide at
100 C, and an initial pressure of 75 bar. Even the slow ramp of
35 bar/min provided a complete separation in approximately five
minutes. As the ramp rates were increased some degradation of
resolution became apparent. However, the analysis time was reduced by more than half. A quantitative comparison of the separation quality as defined by the Trennzahl number of selected alkane
pairs for various ramp rates is given in Table IV. The retention
time (T_R) for one of the alkane components in each pair is also
listed. The final column in the table defines the separation power
per unit time. The Trennzahl numbers decreased by a factor of two
when the ramp rate was increased from 25 to 160 bar/min. However,
the analysis time decreased by a factor of 2.5. Consequently, the
separation power per unit time (TZ/T_R) actually increased with the
more rapid ramp rates. During pressure ramping the linear velocity
of the mobile phase through the column generally doubles when going

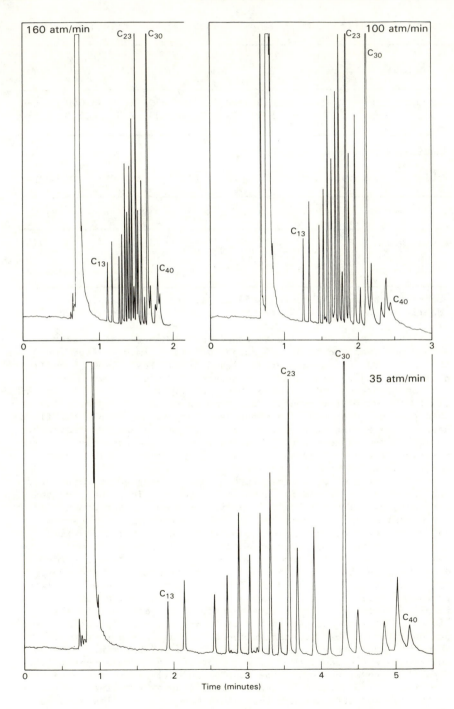

Figure 13. Comparison of three different rapid pressure ramps for the rapid SFC separation of an alkane mixture.

from 75 to 350 bar. Increases of up to 8 cm/sec appear to have negligible effects on the chromatographic performance. In addition, a rapid pressure ramp gives rise to an effective velocity gradient across the column (27). Consequently the average linear velocity is substantially higher than suggested by the initial void volume.

Table IV. Separation Number Comparison for Various Pressure Ramps[a]

Ramp (bar/min)	TZ C_{14}-C_{16}	$T_{R(min)}$ C_{14}	TZ C_{26}-C_{28}	$T_R(min)$ C_{26}	TZ C_{14}-C_{28}	$T_R(min)$ C_{28}	TZ/T_R[b]
25	12.3	2.2	3.4	3.9	49.2	4.1	11.9
35	11.0	2.1	3.6	3.9	50.3	4.1	12.3
50	9.6	1.9	3.2	3.1	40.5	3.3	12.3
75	6.8	1.5	2.5	2.3	31.1	2.4	12.7
100	7.0	1.3	1.8	1.9	30.9	2.0	15.4
160	6.4	1.2	1.2	1.6	25.5	1.6	15.7

[a]Data obtained on a 1.5 m X 25 m i.d. column with a mobile phase linear velocity of 3.5 cm/sec at the initial starting pressure of 75 bar.
[b]Separation power per unit time, TZ/T_R(min) (C_{14}-C_{28}/$T_R C_{28}$).

A potentially important application area for capillary SFC and SFC-MS is in the analysis of thermally labile molecules not readily amenable to gas chromatography, such as mycotoxins of the trichothecene group (48,49). Figure 14 shows a fast separation of four trichothecenes on a short 0.8 m x 25 μm column at 100 C with supercritical CO_2 as the mobile phase. Diacetoxyscirpenol (DAS) and T-2 toxin are easily resolved while the two macrocyclic compounds, roridin A and verricarin J, are not well separated. However, even this level of separation is often sufficient given a highly selective detector such as the mass spectrometer. Application of SFC-MS to these compounds is described in detail elsewhere (42).

The major advantage of capillary SFC for fast separations compared to packed column methods results from the use of pressure programming to rapidly vary the mobile phase solvating power. The capillary SFC methods provide rapid separations and peak widths nearly ideally compatible with typical maximum mass spectrometer scan speeds. Since the total SFC effluent enters the mass spectrometer, quantitation is straightforward. In addition, significantly higher sensitivities can be obtained due to the narrow peak widths and detection limits in the sub-picogram range have been found using selected ion monitoring techniques (9,42).

Mass Spectrometric Methods for Capillary SFC-MS. A significant advantage associated with capillary SFC-MS methods, and in contrast to all mechanical (e.g., moving ribbon) HPLC-MS interfaces, results from the flexibility in selection of ionization methods. Although initial studies were conducted using chemical ionization, and it remains the method of choice for most applications, the DFI process is also compatible with electron impact ionization (37).

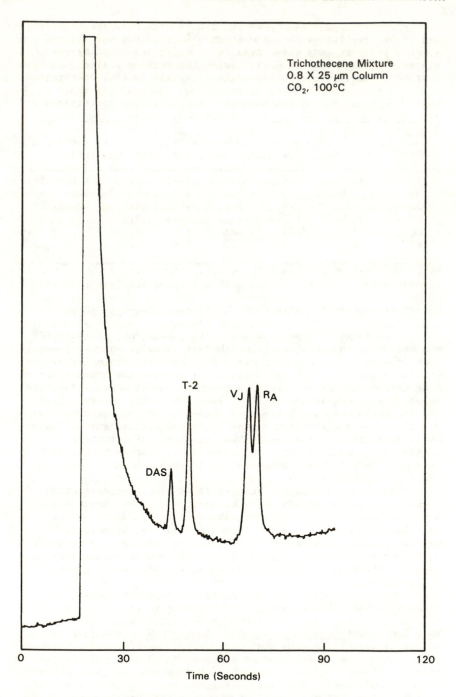

Figure 14. Rapid capillary SFC/MS separation of a standard
mixture of diacetoxyscirpenol (DAS), T-2 toxin and the macrocyclic trichothecenes verricarin J and roridin A.

Electron impact mass spectra also provide more spectral information than typically contained in CI spectra. However, as the
range of SFC-MS application broadens to more polar and higher molecular weight materials chemical ionization methods will become more
valuable. In addition to the higher sensitivity usually obtained
with CI, the identification of an unknown compound obviously benefits from knowledge of the molecular weight typically obtained by
CI. Since many materials of interest will not be in existing
libraries of mass spectra, the value of EI spectra is reduced,
particularly for high molecular weight compounds where few peaks
may be obtained near m/Z values indicative of the molecular species.
Proper selection of CI reagent gases may be the most effective
method for obtaining the combination of molecular weight and structural information for characterization of unknown materials. More
sophisticated mass spectrometric methods (e.g., MS/MS and high
resolution MS) with SFC are already being explored in this laboratory to meet these demands (42,51).

Future Directions

The recent progress in capillary SFC-MS has demonstrated significant potential as an favorable alternative to HPLC-MS for many
applications. The capillary SFC instrumentation is relatively
simple and the practice of SFC with fluids such as CO_2 is amenable
to routine laboratory application. The improved chromatographic
resolution and sensitivity possible with capillary SFC-MS compared
to LC-MS should serve to make this approach increasingly attractive.
    While the potential exists for rapid growth in SFC and SFC-MS
applications, practical problems remain to be resolved. Improved
injection techniques are required and, in particular, on-column
injection techniques which remove sample solvent prior to analysis
and eliminate restrictions due to low sample solubility or limited
injection volumes. The limitations for more polar materials are
mostly unexplored. A large fraction of the compounds presently
separated by HPLC can, in principle, be analyzed using SFC. In
fact, it is not unreasonable to predict that essentially any compound soluble in an organic solvent should be amenable to SFC.
Many highly polar materials are soluble in supercritical fluids
such as ammonia. As an example, Figure 15 gives the mass spectra
of 2-deoxyadenosine obtained using supercritical ammonia for DFI-
MS. At present the most likely classes of compounds to be inaccessible by SFC are those soluble only in aqueous solutions and where
specific chemical interactions (such as hydrogen bonding) are required for solubility.
    Major applications of SFC and SFC-MS will undoubtedly be in
the analysis of high molecular weight polymer and organic mixtures.
The potential for separation of high molecular weight polymers has
been well established. An example showing the application to analysis of the high molecular weight triacylglycerols obtained from a
supercritical fluid extraction of butter is given in Figure 16.
The figure shows major components were separated in less than three
minutes. A pressure ramp of 100 bar/min starting at 75 bar was
utilized, demonstrating the high speed capability of capillary
column separations. Figure 17 shows the mass spectrum of palmitoyl-
disteraroylglycerol (MW = 862) obtained during the capillary SFC

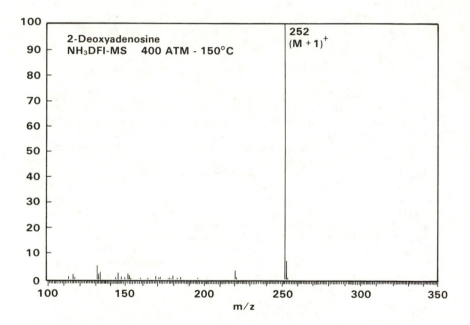

Figure 15. Supercritical ammonia DFI-MS chemical ionization
spectra of 2-deoxyadenosine.

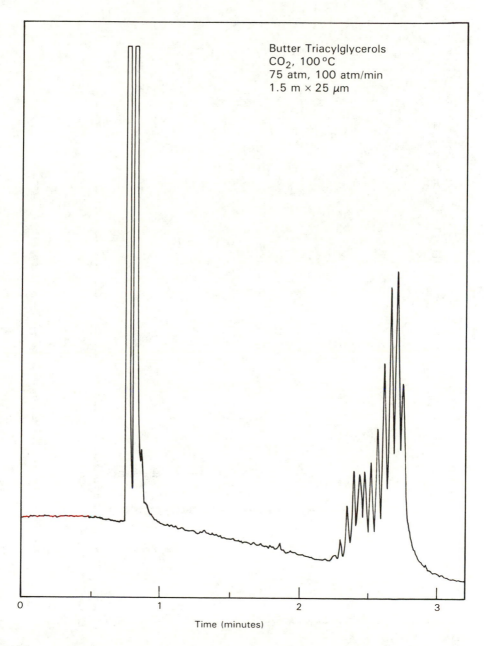

Butter Triacylglycerols
CO_2, 100 °C
75 atm, 100 atm/min
1.5 m × 25 μm

Time (minutes)

Figure 16. Rapid capillary SFC separation of butter triacyl-glycerols.

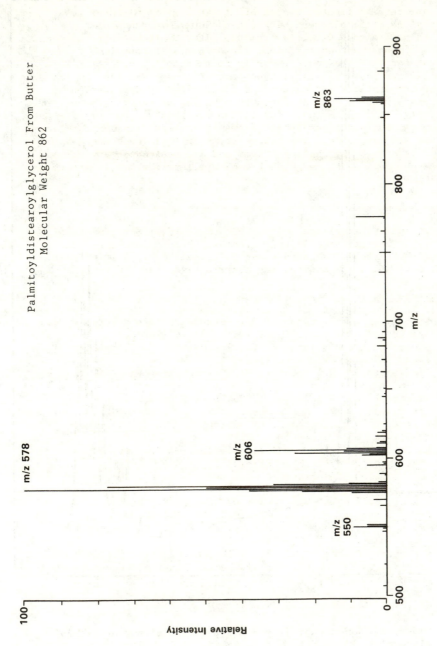

Figure 17. Mass spectra of a triacylglycerol obtained during a
capillary SFC-MS separation (See Figure 16).

separation of this mixture which further illustrates the ability
to characterize compounds of low volatility and high molecular
weight.
     The current limitations of SFC technology result from the
early stage of development of the technique. Current columns have
been adapted almost directly from capillary GC technology. The
demands imposed by supercritical fluids are different from those
encountered in GC and are primarily related to the necessity for
chemical and physical stability of the stationary phase. At present,
higher temperatures and more polar fluids lead to degradation of
stationary phases, causing frequent plugging of restrictor devices.
Similarly, improved methods for column deactivation and formation
of homogeneous stationary phase films in smaller diameter capillary
columns are desirable. Current research in this and other laboratories is addressing these problems and improved capillary columns
allowing a much wider range of SFC-MS applications can be anticipated.

Acknowledgments

We gratefully acknowledge the support of the U.S. Department of
Energy, Office of Basic Energy Sciences, through Contract
DE-AC06-76RLO-1830. We also thank H. T. Kalinoski, C. R. Yonker,
S. L. Frye and A. J. Kopriva for contributions to this work.

Literature Cited

1. Novotny, M.; Springston, S. R.; Peaden, P. A.; Fjeldsted, J.
    C.; Lee, M. L. Anal. Chem. 1981, 53, A407-A414.
2. Fjeldsted, J. C.; Kong, R. C.; Lee, M. L. J. Chromatogr.
    1983, 279, 449-455.
3. Springston, S. R.; Novotny, M. Chromatographia 1981, 14,
    679-684.
4. Peaden, P. A.; Fjeldsted, J. C.; Lee, M. L.; Novotny, M.;
    Springston, S. R. Anal. Chem. 1982, 54, 1090-1093.
5. Wright, B. W.; Udseth, H. R.; Smith, R. D. J. Chromatogr.
    1984, 314, 253-262.
6. Smith, R. D.; Felix, W. D.; Fjeldsted, J. C.; Lee, M. L.
    Anal. Chem. 1982, 54, 1883-1885.
7. Smith, R. D.; Fjeldsted, J. C.; Lee, M. L. J. Chromatogr.
    1982, 247, 231-243.
8. Yonker, C. R.; Wright, B. W.; Udseth, H. R.; Smith, R. D.
    Ber. Bunseges. Phys. Chem., 1984, 88, 908-911.
9. Smith, R. D.; Udseth, H. R.; Wright, B. W.; Kalinoski, H. T.
    Anal. Chem., 1984, 56, 2476-2480.
10. Schneider, G. M.; Stahl, E.; Wilke, G., "Extraction with Super-
    critical Gases", 1980, Verlag Chemie, Deerfield Beach, Florida.
11. Smith, R. D.; Udseth, H. R. Sep. Sci. Tech. 1983, 18, 245-
    252.
12. Smith, R. D.; Udseth, H. R. Biomed. Mass Spectrom. 1977, 10,
    577-580.
13. Smith, R. D.; Udseth, H. R. Anal. Chem. 1983, 55, 2266-2272.
14. Albaugh, E. W.; Borst, D.; Talarico, P. C. "Book of Abstracts"
    185th National Meeting of the American Chemical Society,

Seattle, WA, March 20-25, 1983; American Chemical Society: Washington, DC, 1983; Paper ORPL 135.

15. Giddings, J. C.; Myers, M. N.; McLaren, L.; Keller, R. A. Science 1968, 162, 67-73.

16. Gitterman, M.; Procaccia, I. J. Chem. Phys. 1983, 78, 2648-2652.

17. Myers, M. N.; Giddings, J. C. Sep. Sci. 1966, 1, 761.

18. McLaren, L.; Myers, M. N.; Giddings, J. C. Science 1968, 159, 197-199.

19. Klesper, E. A.; Corwin, H.; Turner, D. A. J. Org. Chem. 1962, 27, 700-701.

20. Sie, S. T.; Rijnders, G. W. A. Sep. Sci. 1967, 2, 699-727.

21. Sie, S. T.; Rijnders, G. W. A. Sep. Sci. 1967, 2, 729-753.

22. Sie, S. T.; Rijnders, G. W. A. Sep. Sci. 1967, 2, 755-777.

23. Giddings, J. C.; Myers, M. N.; King, J. W. J. Chromatogr. Sci. 1969, 7, 276-283.

24. Gouw, T. H.; Jentoft, R. E. Advan. Chromatogr. 1975, 13, 1-40.

25. Sie, S. T.; Van Beersum, W.; Rijnders, G. W. A. Sep. Sci. 1966, 1, 459-590.

26. Jentoft, R. E.; Gouw, T. H. J. Chromatogr. Sci. 1970, 8, 138-142.

27 Smith, R. D.; Wright, B. W. Anal. Chem., submitted.

28. Vestel, M. L. Mass Spectrom. Rev. 1983, 2, 447-480.

29. Milne, T. A. Int. J. Mass Spectrom. Ion Phys. 1969, 3, 153-155.

30. Giddings, J. C.; Myers, M. N.; Wahrhaftig, A. L. Int. J. Mass Spectrom. Ion Phys. 1970, 4, 9-20.

31. Randall, L. G.; Wahrhaftig, A. L. Anal. Chem. 1978, 50, 1705-1707.

32. Randall, L. G., Ph.D. Thesis, University of Utah, 1979.

33. Randall, L. G.; Wahrhaftig, A. L. Rev. Sci. Instrum. 1981, 52, 1283-1295.

34. Gouw, T. H.; Jentoft, R. E.; Gallegos, E. J. 6th High-Pressure Sci. Technol. AIRAPT Conf. 1979, 583-592.

35. Henion, J.; "Microcolumn High-Performance Liquid Chromatography", J. of Chromatog. Lab. 1984, 28, Elsevier (P. Kucere, Ed.), Amsterdam, Chapter 8, 260-300.

36. Ashkenas, J.; Sherman, F. S. "Proceedings of the 4th International Symposium on Rarefied Gas Dynamics", 1966; deLeevw, J. H., Ed.; Academic Press: New York, 1966.

37. Hagena, O. F.; Obert, W. J. Chem. Phys. 1972, 56, 1793-1802.

38. Smith, R. D.; Udseth, H. R.; Kalinoski, H. T. Anal. Chem., 1984, 56, 2971-2973.

39. Fjeldsted, J. C.; Kong, R. C.; Richter, B. E.; Fields, S. M.; Jackson, W. P.; Lee, M. L. 1984, Pittsburgh Conf. on Anal. Chem. and Appl. Spectroscopy, Atlantic City, NJ, March 5-9, Paper No. 596.

40. Peaden, P. A.; Wright, B. W.; Lee, M. L. Chromatographia 1982, 15, 335-340.

41. Wright, B. W.; Peaden, P. A.; Lee, M. L.; Stark, T. J. J. Chromatogr. 1982, 248, 17-34.

42. Smith, R. D.; Wright, B. W.; Udseth, H. R. J. Chromatogr. Sci., in press.

43. Gere, D. R.; Board, R.; McManigill, D. Anal. Chem. 1982, 54, 736-740.
44. Guiochon, G.; Colin, H. "Microcolumn High-Performance Liquid Chromatography", (Ed. P. Kucera) Elsevier, New York, 1984; Chapter 1.
45. Novotny, M.; Springston, S. R. J. Chromatogr. 1983, 279, 417-422.
46. Peaden, P. A.; Lee, M. L. J. Chromatogr. 1983, 259, 1-16.
47. Yost, R. A.; Enke, C. G. J. Amer. Chem. Soc. 1978, 100, 2274-2275.
48. Bata, A.; Vanyi, A.; Lasztity, R.; Galacz, J. J. Chromatogr. 1984, 286, 357-362.
49. Rosen, R. T.; Rosen, J. D. J. Chromatogr. 1984, 283, 223-230.
50. Dorough, H. W.; Thorstenson, J. H. J. Chromatogr. Sci. 1975, 13, 212-224.
51. Chess, E. K.; Udseth, H. R.; Smith, R. D. Anal. Chem. in press.

RECEIVED April 19, 1985

Author Index

Subject Index

T

U

Production by Hilary Kanter
Indexing by Keith B. Belton
Jacket design by Pamela Lewis

Elements typeset by Hot Type Ltd., Washington, DC
Printed and bound by Maple Press Co., York, PA